WHAT PEOPLE A

A DIET OF AUSTERITY

Elaine has a produced a must-read book for all of us concerned with combatting climate change. We can't diet our way to a better world but we can and must change the system to sustain the future. This book is well written, fascinating, controversial and essential.
Derek Wall, former Green Party of England and Wales Principal Speaker and author of *The Rise of the Green Left*

Who is to blame for climate change? Graham-Leigh says it's not fat people, cows or the working class. A challenging and interesting book, packed with new ideas to make you think again about what you thought you knew.
Jonathan Neale, author of *Stop Global Warming, Change the World*

Food production and consumption is increasingly recognized as a major driver of global climate change, but where does the balance of blame lie? Elaine Graham-Leigh's book makes an important and timely contribution to the debates swirling around food production systems, 'overconsumption' and climate change. The class-based ideological assault on working people as the culprits of our multi-faceted ecological crisis are elucidated with great clarity by the author. For people who believe combining social justice with ecological sustainability is an essential part of the answer, Graham-Leigh's book provides essential ammunition as to why it is our system of food production for commodity exchange, and the waste inherent in it, rather than individual behavior and food choices, that lies at the heart of the crisis.
Chris Williams, author of *Ecology and Socialism*

A Diet of Austerity

An excellent, polemical book; enjoyable, interesting and very, very well-written. A really new contribution to theory and an incisive argument.
Chris Nineham, author of *The People v Tony Blair*

A Diet of Austerity

Class, Food and Climate Change

A Diet of Austerity

Class, Food and Climate Change

Elaine Graham-Leigh

Winchester, UK
Washington, USA

First published by Zero Books, 2015
Zero Books is an imprint of John Hunt Publishing Ltd., Laurel House, Station Approach,
Alresford, Hants, SO24 9JH, UK
office1@jhpbooks.net
www.johnhuntpublishing.com
www.zero-books.net

For distributor details and how to order please visit the 'Ordering' section on our website.

ISBN: 978 1 78279 740 1
Library of Congress Control Number: 2014949960

A CIP catalogue record for this book is available from the British Library.

Design: Stuart Davies

Printed in the USA by Edwards Brothers Malloy

CONTENTS

For Dominic Alexander

Preface

Like any book, this project has had many beginnings, but its real origin was perhaps the Campaign Against Climate Change demonstration in London in late 2008.

The national climate change marches, timed to coincide every year with the international climate change talks, had been held since 2005. Whatever the official focus of the demonstration they always showed a wide mix of placards, with people advocating every green position from direct action to beekeeping. I wouldn't have expected the 2008 demonstration to be an exception, but what did strike me, looking around the protestors gathered in Parliament Square, was the prevalence of placards about food; in particular about meat and its contribution to climate change.

I hadn't considered food's role in climate change since I was first in the movement and argued with older activists about the greenness or otherwise of eating chips at the bus stop on the way home from meetings. I felt a little blindsided, as if I had missed a memo. I also felt that there was a disconnect somewhere in using such a collective statement of people power as a national demo to call for an action like giving up meat, which could only be done by individuals at home.

Sitting on one of the handy walls around the edge of the square, I got into conversation about the day with the woman next to me. Despite the decent turnout, she was in a pessimistic mood. People just weren't prepared to change their behaviour, she told me. We would never persuade them to give up their cars, and all the campaigning in the world for better public transport wouldn't change that. However, she added, brightening up, what we could do was persuade everyone to give up eating meat. That was much more realistically achievable and would be the best thing we could do for the climate. I didn't think I agreed, but couldn't muster any more coherent arguments than that I personally found the idea of changing my travel habits easier and less intrusive than my eating patterns. As she pressed a Vegan Society leaflet on me, I came to two decisions: that I would wait until I was in the tube to eat the pork pie I had in my packed lunch, and that this food argument needed looking into. This book, after many changes of direction along the way, is the result.

Thanks are due to many people: to everyone who took time out of their day to let me come and ask them questions; to Charlotte Cooper for posting about the project on her site and finding me more people to pester; to my comrades in Counterfire for support, engagement and political argument. I am particularly indebted to Chris Nineham for perceptive comments on an early draft and to Feyzi Ismail for encouragement and friendship. Angela Graham-Leigh generously put aside our political differences for long enough to proof-read and was glutton enough for punishment to offer to do it twice. Lastly, without Dominic Alexander's insights, support and belief in the project this book would never have been started, let alone finished. In gratitude for his love and comradeship, it is dedicated to him.

Introduction

It is a familiar and depressing story: the poor blamed for poverty, and more, for taking resources from everybody else.

When a £500-a-week benefit cap was introduced in April 2013 by Haringey Council in North London, one of the first local authorities in the UK to try it, 'Susan' and 'Samantha' decided to step forward and let the local benefit justice campaign highlight their cases. The details were shocking enough, showing the hardship people across the borough would be facing. For both, single mothers with a number of children, the effect of the cap was to leave them having to find money for their housing costs out of their already tight budgets. As the benefit justice campaign explained to its supporters on its Facebook page, Susan, who has seven children, used to get £537 in benefits per week, plus £245 in housing benefit. The £500 cap took £282 away and left her with £245-worth of rent to find out of her £500 maximum benefits, and so £255 for everything else. That's well below a minimum decent standard of living (the Joseph Rowntree Foundation estimates that to get this, a single parent with three children would need £458 a week) and between eight people, works out at under £32 each for the week.

With facts like these, Susan and Samantha could have expected a sympathetic hearing from the campaign's Facebook followers. Some did indeed agree that this was a disgraceful way for a rich country to treat its poorest citizens, but there were plenty of opposing views. 'Does this woman with 4 kids and no income think that life is any different for those of us that fund her extravagant lifestyle choices?' asked one commenter about Samantha.

> When there's less money around we all have to make sacrifices. I do two jobs to keep my family afloat. I also have to pay

> towards the upkeep of her family. Sorry, but I want my family to come first. I work. Why did she have 4 kids when she has no income with which to support them? I am totally gobsmacked at her insistence that she must stay in one of the world's most expensive cities at our expense, while we all have to make more and more sacrifices to fund her extravagances.

The verdict on Susan, with her large family, was even worse (all spelling and ellipses in originals).

> I have to say I believe that having seven children when you are on a low income is totally irresponsible…It does not take an Einstein to work out that when your house is getting full you STOP having more children.

*

> So why do you have a kid with a complete loser Susan? Responsibility is a two way street...People should put the children first, but the welfare system was not devised for large families with no income parents, that's why we need a social security system that actually logs what the individual pays in, thus creating a culture of wanting to work....the more you pay in...the more security you gain. There should be a minimum level of support...but we need a system of fairness, not just...entitlement.

*

> Why should tax payers pay for someone to have 7 children and has never contributed in her life! Sounds like another sponger that thought she could pop out kids left right and centre and expect to be paid for! Back fired slightly! Good let

> it be a lesson to those with the same attitude that you can't pop them out and expect everyone else to pay for you!!!
>
> *
>
> If your on the dole and you cannot afford to have kids then u shouldnt be able to churn them out like no tomorrow.... Something needs to be done.... Up and down the country there are people churning them out just to get more money because they beleave the state owes them something, yet we see these people with drink, fags, tattoos etc etc.... Knock all that shite on the head for a start....
>
> *
>
> She should have kept her fucking legs together and got a job people like this make me sick.

Unsurprisingly, both women were shocked by the vituperative response they received, and neither is now prepared to raise their heads above the parapet again. In part, of course, this story is simply illustrative of man's web-based inhumanity to man, but it has also a wider significance as a demonstration of how far the idea of the deserving and the undeserving poor has got into people's heads. It seems to be a common view that, as one of the Facebook commenters said:

> Those who maintain unhealthy lifestyle choices and do not seek to reign anything in, are the ones who do nothing to support the original arguement. Those who genuinely need the support, who have fallen on hard times, and are willing to contribute a bit of effort, without hesitation are the ones who deserve to be supported. The rest need to understand what 'responsibility' is!

The 'undeserving' here are those who are perceived not to be willing to work, as Susan and Samantha were charged with being, but also those who have 'unhealthy lifestyle choices' through insufficient self-restraint; in other words, who eat and/or drink too much.

That benefit recipients are effectively being told to get off their fat behinds and work can seem so unremarkable that it's easy to overlook its significance. There is a line from the crude abuse levelled at Susan and Samantha, through government austerity programmes, to a growing view of who is to blame for climate change and global hunger. In the years since the collapse of Lehman Brothers in 2008 ushered in the economic crisis, governments across the Western world have attempted to convince us that the problems have been caused by overspending, and that now we all have to sacrifice in order to get the economy back on track. Phrases like 'tighten our belts' and 'balance the budget' have been commonplace. The idea is to make us equate the national budget with our personal finances, underlining the notion that the crisis of the system was the fault of individual consumers on a spending spree. This attempts to make the austerity agenda understandable: if 'we' have all overspent, it follows that 'we' have to cut back, whether that's accepting job losses and pay freezes, sacrificing libraries or arts funding, or denying those on benefits luxuries like more bedrooms than the government thinks they need.

The hypocrisy of a government of multi-millionaires telling us that 'we are all in this together' is evident, but the argument was not developed solely to allow Tories to pose as men of the people. The idea that crises of the system are caused by the self-indulgence of individuals, and therefore can be solved by those individuals just changing their behaviour, has a long pedigree in green issues. In this sense, those who attempt to blame economic crises on people who are perceived to have chosen to be a burden on the system are using a way of thinking which is already common in discussions about climate change. This is a view of climate change which sees it as arising primarily from individual behaviour; people 'choosing' high-emission lifestyles who therefore need to be cajoled out of them like people 'choosing' to have children while living on benefits.

The shelves of books advising people on how to reduce their personal carbon emissions might give the impression that this view of the causes of environmental destruction is hegemonic, but in fact its value has always been contested. From a tactical perspective, an influential report, *Weathercocks and Signposts,* issued by the World Wildlife Fund (WWF) in 2008, argued that making small changes might not lead most people to go on to make more significant alterations to their lifestyles. For most, it suggested, the small changes might be all that they would do before they stopped, feeling that they had 'done their bit'.[1] Others have pointed out that while the ideology of capitalism maintains that the consumer is all-powerful, this is not necessarily true in practice: individuals have less power to change their own lifestyles, let alone the system in which they live, than these arguments would have it. With an interesting synchronicity, however, this view of ecological crisis as the product and responsibility of individual lifestyles was given new impetus just at the time that the economic system was lurching into crisis.

In September 2008, as Lehman Brothers were filing for bankruptcy, we learned that climate change was largely caused by that most individual of consumption choices: what we eat. Dr Rajendra Pachauri, chair of the UN Intergovernmental Panel on Climate Change (IPCC), argued in a lecture that livestock farming was such a problem for the climate that everyone should have one meat-free day a week, as a step towards cutting their meat consumption even further.[2] These comments were essentially echoed by other authorities and were widely reported in the press, with headlines like 'UN says eat less meat to curb global warming',[3] 'Meat must be rationed to four portions a week'[4] and 'Government advisor: eat less meat to tackle climate change'.[5] At the same time, the Food Climate Research Network issued what would prove to be a very influential report arguing that the food system in the UK was responsible for 19% of our greenhouse gas emissions. This was against the background of dramatic global food price rises, which led to demonstrations and riots in countries from El Salvador to Indonesia as people protested at the cost of staple goods soaring beyond their reach. Undoubtedly, this contributed to a general sense of a food crisis; the idea that the world had arrived at the point where environmental depredation had finally meant that we lacked the ability to feed everyone.

The effect of this was to place the idea of food production as a significant problem, if not the most significant contribution to ecological crisis, firmly within the mainstream. It has not proved a flash in the pan; that food is a major climate change issue is now a given in much of the media. It's difficult to over-stress just how marked a shift, in such a short time, this is, elevating one area of modern production to centre-stage from its previous peripheral role. So minor was food's contribution to climate change previously considered to be, compared to areas like power generation or transport, that comprehensive works like George Monbiot's *Heat* (2006)[6] or Jonathan Neale's *Stop Global Warming. Change the World* (2008)[7] scarcely accord it a mention. Even a work devoted to individual lifestyle changes, like Chris Goodall's *How to Live a Low-Carbon Life* (2007), keeps the question of emissions reductions from dietary changes to sixteen pages, after the chapters on heating, lighting, household appliances, cars, public transport and aviation.[8]

In many ways, of course, the addition of food production to our understanding of climate change problems is both good and inevitable. As scientists gain more and more understanding of how the immensely complex climate mechanisms work, and how human activity is affecting them, what we have to do in mitigation and minimisation will change. Ultimately, however, climate change is a political problem as much as it is a scientific and technical one. The task is not simply to identify how we are damaging the climate, or to invent technical fixes for it, but to make structural changes which will enable us to implement specific solutions and construct a different relationship with the natural world. This is an intensely political challenge, and it means that new revelations, like that of the contribution of food to greenhouse gas emissions, are not simply objective, scientific facts but arguments with significance for the sort of solutions for climate change we call for and the sort of society we want to build.

In the first place, we have to be aware that food is a morally-loaded category in a way not shared by any other climate change villain. Considerations of the food system are replete with comments like this one from the Soil Association's 2010 report on the food crisis: 'Eating is a primal, physical act, but it is also a moral one. It reflects who we are – our character, our values and our ethics'.[9] Aside from the climate implications, there is little morality associated with power use or even with car use: we don't judge anyone's worth according to how they heat their house, and do so according to their SUV ownership principally for climate change reasons. However, what we eat is associated with ideas about our value as people. Eating healthily, however that is understood, is frequently presented as a moral good, as an example of virtuous, responsible behaviour. The concomitant is that eating unhealthily is perceived as a moral failure, along with obesity as visual evidence of this guilt. The understanding that different foods make a significant contribution to climate change fits into this already-established moral framework for eating, so that, as the BBC News website once had it, 'Obese blamed for the world's ills'.[10] Proponents of dietary change as a response to climate change are clearly aware of this: in 2009, for example,

Lord Stern predicted that meat-eating would become morally unacceptable because of its effect on the climate, in the same way that drink-driving was once regarded as a foible and has now become a crime.[11]

Concerns about obesity coupled with the moral loading of food choices mean that the role of government and other authorities in hectoring people about diet is well-established. Outside of climate concerns we do not expect to be lectured about our transport choices, but we are told what we should and shouldn't eat on a daily basis. One immediate effect of this is that a greater prominence of food in climate change campaigning casts individual lifestyle changes, as opposed to changes at the level of the system, as most important. So in the 2008 coverage of the food and climate change issue, it was clear that if food production was the problem, a dietary shift away from meat was the solution.

To a degree, it could be argued that this was a media creation. Media coverage will seize on the most populist angle, and it is not surprising that the press would think that the idea that people would be expected to change what they eat would sell more papers than a headline about reform of farming practices. But the concentration on dietary shifts did emanate from the original material, however much it was seized on in the media coverage. Indeed, in November 2008, Rajendra Pachauri teamed up with Paul McCartney to write to *The Independent* that becoming vegetarian would be 'the single most effective act' that anyone could take to reduce greenhouse gas emissions.[12]

The discovery of diet as a cause of climate change gives a fillip to the notion of sacrifice for the sake of the climate. It is probably fair to say that the notion that we will have to sacrifice our Western living standards to address climate change is a truism for much of the green movement, whether this is achieved through individuals choosing to change their lifestyles or through system change in which such sacrifices would be mandated. The concept of different foods as sinful pleasures is so ingrained into our thinking about food in general that the conclusion that we will all have to sacrifice our current diets seems to have been an easy one to reach. It's notable that in much of the discussion of food and its impact on the climate, what is characterised as a typical Western diet (high in fat and sugar, lots of junk food, lots of red meat) is viewed as both uniquely bad for the climate and uniquely desirable. There is an underlying assumption that everyone in the world would eat this way if they were able to; but such is its power to damage the climate that actually, no one should.

This matters because the notion of sacrifice goes to the heart of the matter. There is a long-standing tradition, going back to the post-Second World War period, of identifying Western overconsumption as a serious problem for the world, even before the reality of climate change was understood. The idea of sacrifice in climate change is often seen in terms of luxury consumption driven by advertising – yes, we don't want to immiserate people, but we all buy high-end electrical goods that we don't really need, right? However, an analysis of how food works within the overconsumption discussion reveals that it has a distinct effect on the class identification of those who are deemed to be overconsuming.

Another extremely important development in campaigning on climate change has been the emergence of the idea of climate justice. The idea that Westerners have to sacrifice to stop climate change is often seen as part of this justice agenda, on the grounds that we are taking more than our fair share of the world's resources and leaving the poorest people in the world to deal with the consequences of our consumption. There is a lot in this argument, and I would not for one moment wish to minimise the sincere commitment of people in the green movement to justice and fairness. However, the question of who sacrifices and what they have to sacrifice is a key one. David Cameron's 'we're all in this together' hides an austerity agenda deliberately focused on attacking the poorest and most vulnerable in order to protect the wealth of the richest. In the same way, the notion that everyone in the rich West is guilty of overconsumption and can comfortably sacrifice conceals enormous differences between those promulgating the argument and those who are being expected to change. The necessity of tackling climate change can be used to justify attacks on ordinary people, if it's their consumption which is defined as problematic. Climate justice and sacrifice, as revealed by the food issue, may not be compatible.

At a time when working people are under attack from the government's austerity agenda, the danger is real that casting food choices as a major cause of climate change provides another layer of justification for cuts. The focus on food as an issue for climate change must be seen in the context of longstanding arguments about the malign effects of overconsumption, both on the planet and on individuals, which have themselves become increasingly focused on edible as opposed to luxury consumption. This has the effect of shifting the identification of the overconsumers from the rich to the working class, since obesity is overwhelmingly identified as a problem of the poorest people in the West. We are back, it seems, with Malthus' idea that the problems of the world are caused by the appetites of the poor.

Part of the Malthusian view of the world is the fear that we are at the limit of the planet's carrying capacity: that whether because of greed inherent in human nature or simply because of numbers, there are now too many people for us all to be fed. On the contrary, I argue that it is capitalism, not the human population, which is straining the natural limits of production. The situation with capitalist production tells us very little about what production under a different system might look like, but we can be confident that without the wastefulness and inequality which are part of capitalism's makeup, our ability to feed the world's population decently and fairly would be greatly increased. It is not particularly helpful, however, to opine about how much better off we would be in a non-capitalist system without some suggestion about how we might get there. I conclude therefore with some consideration of the environmental movement and how people who care about climate change and justice can best fight back against the system we face.

1

Working class blamed for world's ills

There is one statistic which for many people sums up everything that is wrong with the West and our food consumption: that across the world, a billion people are starving while another billion are overweight. As might be expected for such a clear and popular idea, the figures change slightly depending on the source: Raj Patel, for example, begins his popular *Stuffed and Starved* with 'the hunger of the 800 million' and 'another historical first: that they are outnumbered by the one billion people on this planet who are overweight',[13] while in some recent iterations, the numbers of the overconsumers have increased to two billion.[14] Regardless of the precise scale, the idea of the billion versus the billion is so widely repeated that it appears an almost compulsory beginning for commentary on the problems of the modern food system.[15]

The striking juxtaposition contains a number of assumptions, which can be more or less clear but nevertheless are always there. The first key assumption is that food supplies, whether as a result of climate change and environmental degradation, or simply because the planet is finite, are sufficient but limited. There is enough for everyone but for everyone who overconsumes, someone else has to get by on less to make up for it. This was expressed clearly in 1999 by Thomas Princen in one of the earliest versions of the idea: 'the overconsumption of the billion or so who consume far more than their basic needs and, it is reasonable to assume, contribute directly or indirectly to the underconsumption of the impoverished billion'.[16] Princen was not necessarily only thinking about food here (this will be discussed further in the next chapter), but later iterations are clearly considering eating and drinking to the exclusion of other sorts of consumption. Patel's reference to a billion people being overweight makes clear that the overconsumption under discussion is overconsumption of food. This is not a lone example. The Commission on Sustainable Agriculture and Climate Change, for example, talked in their 2011 report on food security about overconsumption in general in their introductory remarks, but substantiated this on the following page with a table showing 0.9 billion undernourished people in the world and 1.5 billion people over the age of twenty who were overweight.[17]

The second assumption is that this is a matter of individual behaviour. In other words, if the developed world is overconsuming food, to the detriment of the developing world, this is caused by, and essentially the same as, individuals overconsuming. Thus, for example, a 2008 study on reducing fossil fuel inputs in food came to the decided – and according to EScience News 'very astute' – conclusion that the most important thing that individuals could do to help deal with problems of food and climate change would be to eat less.[18] Since the article used the figure for the US food production per head (here 3,747 calories) as the amount of food consumed, this is perhaps not surprising, although it is worth noting that not all food produced is necessarily eaten.[19] On one level this may seem obvious – no one, after all, is being force-fed – but in fact it goes to one of the core beliefs of neo-liberal economics: that the motor of everything in society is the choices made by consumers in the market. The problems of food production, and inequalities of food distribution, are caused ultimately by the purchasing decisions of individuals, and the companies involved would change their ways if those consumers demanded differently.

Since it is obvious that individual consumers do not have a completely free choice of what they spend their money on, uninfluenced by the persuasive efforts of the businesses hoping to receive their custom, this has become the theory of the obesegenic society. On one side, this is a recognition that there are factors other than greed behind individuals' food choices. A 2013 version of the billion versus the billion, for example, contrasted 'the thin and the fat' as exemplified by a Pakistani peasant farmer and a Canadian university student who put on weight because of the limited options available to her in her college cafeteria.[20] Other criticisms of Western eating practices highlight the efforts of food manufacturers and supermarkets to entice us into buying food we don't 'really need',[21] or simply blame the sheer availability of food. For some, it is apparently 'the availability of relatively inexpensive and highly palatable foods in almost unlimited abundance' which leads 'affected individuals [to] eat many times a day and consume large portions'.[22] 'Overconsumption of food' apparently 'is part and parcel of a society in which consumption and consuming is its raison d'être'.[23]

These insights do not, however, remove the responsibility from individuals for their food consumption. It is noticeable that the same writers who identify the efforts of the food industry to maximise sales as important are also those whose main conclusions are around ways to eat less, whether that is advice to switch to a daily shopping trip on foot with a rucksack[24] or calls for a paradigm shift in our food preferences, 'a form of personal perestroika'.[25] The assumption that issues of food come down to individual food consumption choices remains paramount, and understanding the ways in which the system may encourage fatness only gives greater urgency to the message that individuals must make their own efforts to become thin. Indeed, the obesegenic environment itself has become for some another stick to beat fat people with, as shown particularly starkly in a well-known 2007 article called 'Fat Bastards', in which fat people became proxies for all aspects of Western overconsumption and appropriation of global resources: 'living metaphors for the way the United States is viewed by much of the rest of the planet: a rapacious, gluttonous, insatiable nation of swine… wallowing in the mud of our laziness and indifference'.[26]

This leads neatly on to the third assumption underlying discussions of problems of the global food system: that if it comes down to an issue of individual food consumption, then we can tell who the guilty parties are, since they are the ones who are fat. The elision of the difference between identifying an obesegenic environment and blaming fat people, and only fat people, for the problems caused by the factors which create it is such an obvious one that it is often possible to wonder if in some arguments it isn't subconscious. It does however make a significant difference. An argument which says that the production of large amounts of nutrient-poor, energy-dense food in the West is problematic for food consumption worldwide and for the climate (and has a tendency to make some individuals become fatter than they would otherwise be) is a world away from one which says that regardless of the interests vested in that pattern of food production and consumption, the responsibility lies only with those people who become fat because of it.

'Obese blamed for the world's ills'[27]

That fat people are taking more than their share of scarce resources, to the detriment of everyone else, is a familiar idea. We are, after all, routinely told that obesity in the UK is costing the NHS billions of pounds a year and threatens to bankrupt it, in the same way that the NHS is portrayed as under attack from other undesirable, undeserving groups, like foreign tourists or immigrants from Romania. For anyone whom others identify as fat, the assumption that fat people are a dead weight dragging down the rest of the population is depressingly, constantly evident. Different theories about why people are fat come and go, but the base assumptions – that the aim should be for everyone to be normatively slender, that to be fat is the worst thing in the world, and that if you are so reprehensible as to allow yourself to be fat, you have nothing to contribute – remain the same.

Charlotte Cooper has been a fat-acceptance activist for 25 years[28] and has seen waves of attention to obesity ebb and flow. Obesity, she points out, always makes a good story for journalists to pursue because it allows their readers to feel safely sanctimonious. It sells papers and lets the journalists get paid. Behind the media attention, there are whole industries devoted to weight loss and an academic discipline where fat people are discussed largely in their absence. The idea that fat people could be a social group with interesting things to say about their own experience would be, to most, bizarre. The 'headless fatty' style of image for articles on obesity may have developed as a way of using fat people's pictures without making them easily identifiable, but for Charlotte it's a visual demonstration of how despite all the hysteria about their existence, fat people themselves are silenced.

Talking over coffee in her East London flat, the transformational, enriching, life-affirming possibilities of fat-acceptance activism seem very real, but so also is the ever-present stigmatisation of fat people. We keep coming back to how, in the climate created by the obesity epidemic panic, even people who would consider themselves progressive are happy to share Facebook images which juxtapose fat, black, young children eating McDonald's against starving African children, or send fat-acceptance bloggers infographics suggesting that fat people on treadmills could be used to generate renewable energy. Charlotte's seen too many images like these 'circulated by people who are right on'. She tells me how she's recently written to *The Guardian* to remonstrate about their repeated use of one particularly egregious 'headless fatty' picture. As soon as she starts to describe it I realise I've seen it too: a fat young woman, pushing a pram, walking away, 'Golddigger' emblazoned on her sweatpant-covered behind. 'The scare embodied in that image',

Charlotte says, 'just blows my mind', and she's right. When we look at that picture, we aren't supposed simply to see a person of larger than average size. We're supposed to see not just a fat woman, but a working-class woman, a single mother, a benefits scrounger, a drain on the welfare state, a destroyer of the planet. A picture can indeed say a thousand words.

What *The Guardian*'s image makes clear is the extent to which obesity, supposedly a product of the obesegenic environment to which our very wealth makes us vulnerable, is in fact correlated with lower social class. Fat, it seems, is a very proletarian thing to be. For Charlotte, this identification of fat with the working class was a reality from an early age. Her mother, a nurse, saw Charlotte as the fat one in the family, in need of monitoring and dieting, and her determination to control her daughter's weight came as much from class as from medical concerns. As a working-class family attempting to rise into the middle class, 'it was exposing to them to have a child who was fat'. Charlotte discovered punk and feminism when she was a young teenager (it was helpful that the bookshops where she did most of her reading shelved the feminist books next to the books she was consulting to satisfy her adolescent curiosity about sex), and realised that there was a queer scene she could be part of and that being 'normal' wasn't the be all and end all, but the memory of the stigma is clear.[29]

This is not only a UK experience, nor even one restricted to the English-speaking world. Nicole grew up and still lives in Recife, in Pernambuco in north-eastern Brazil. It's a poorer part of the country, far removed from famous cities of the south-east like Rio de Janeiro and São Paulo, but it is still part of the renowned, body-conscious Brazilian culture, in which women, especially young women, have a responsibility to be slim and sexy and generally to imitate 'The Girl from Ipanema' as far as possible. Nicole says:

> I have been fat since I can possibly remember. Tales of me being a huge baby and a great eater have been part of family folklore. I have a picture of myself as a baby, sitting on top of a table, surrounded by food and eating a whole leg of a turkey while being watched over by the whole family, all smiling and laughing at that funny baby… compared to my little sister (who had always been considered a picky eater), I was a blessing, I ate whatever I was offered.

What was an attractive trait in a baby was less attractive however in a growing girl, and by the time she was thirteen, Nicole was left in no doubt that it was her responsibility to get and remain thin so that she would be desirable to men. In Brazil, it is assumed that fat women in particular are undeserving of male attention, so they are seen as lonely, but also as lazy: 'fat people are to blame for not having enough will power and strength to keep themselves in shape'.[30]

Nicole's family, while not wealthy, are middle-class, so in her childhood having enough money to eat was not the issue it is for many Brazilians. It is noteworthy however that the criticisms of laziness and irresponsibility levelled at working-class people are also again made of people who are perceived to be fat. In Brazil, indeed, obesity levels have recently been blamed explicitly on the poor, as some have suggested that President Lula's 2003 Zero Hunger initiative, which gave poor families the equivalent of just under £15 a month to buy food, may have created a worse public health problem than malnutrition, as the recipients have spent it on cheap, fattening foods rather than on vegetables.[31] The fecklessness of this obesity-causing behaviour is clearly supposed to be apparent. The connection between obesity and the class of those who are obese is not spelled out, but it is there. As with discussions of austerity, it seems that those who have the least are the ones who have the greatest responsibility to be restrained in consuming it.

Obesity and class

There are indications from studies in both the UK and the US that on a population level, obesity may be correlated with poverty. This is not anything so crude as 'all poor people are fat', or even 'all fat people are poor', but when looking at whole districts, areas with higher levels of poverty are more likely than wealthier areas to have higher numbers of fat people. A study by the London Health Observatory of children at ages five and eleven found, for example, that the prevalence of 'children at risk from obesity' was higher the more deprived the area from which the children came.[32] The study did not specify how the risk of children developing obesity was assessed, but the indication of a connection between poverty and obesity found in this study has been repeated by others. The Foresight report, for example, found that 10% more working-class men were obese than upper-class men, while the difference for women was 15%.[33] It is also significant that one of the towns recently designated as the US's fattest – Huntingdon, West Virginia – is also notable for its unemployment and poverty. Situated in Appalachia, the area whose poverty motivated Galbraith to write *The Affluent Society,* it also tops the poll for various other indicators of poverty. More than half of Huntingdon's elderly population, for example, have also lost all of their teeth.[34]

The correlation indicated by these studies between class and obesity may not be generally accepted within obesity science, and this is not necessarily a mainstream view. However, what is important for an understanding of the role of food and obesity within the overconsumption and climate change arguments is not only the as yet fairly limited studies suggesting an actual link between obesity and poverty, but also the much more widespread assumption that obesity is identified with the working class.

As discussed, the contemporary Western media is obsessed

with obesity, and the numbers of illustrations of obesity are so vast that it is only possible to assess a snapshot. However, the images used to illustrate news stories about obesity are likely to play a significant role in general ideas in society about who the obese are and from what class they are likely to come. In consequence, a review of even a limited proportion of these images is worthwhile in assessing the messages which are being sent by common visual representations of obesity. I reviewed the top 150 obesity-related news stories illustrated with images of people on the BBC News website on one day.[35] For the purposes of the study, I excluded news stories which either did not have illustrations or used inanimate images, usually of plates of 'fattening' food, although the latter could clearly tell their own stories about the class presentation of obesity.

45% of the images were of the 'headless fatty' type, with 18% being images of unidentified thinner people (with heads or without) and the remainder of people, fat or otherwise, who were identified by name in the story. It was also notable how limited the stock of anonymous images clearly is. Perhaps indicative of the pressures on journalists to get their stories up on the website as quickly as possible,[36] in proportion to the number of stories requiring illustration, the BBC website appears to have a relatively small number of pictures of anonymous fat people, which it uses again and again. However, even within these constraints, it clearly would be possible for the BBC website's coverage of obesity to tell a number of different visual stories about the sort of people who are obese. The content of the news stories themselves did not tend to give any indication of a class identification of obesity, so it would have been in theory open for the images to attribute obesity to any, or indeed all, social groups.

While portrayals of class may be subjective, a consideration of some putative examples shows that class can indeed be indicated visually. If news stories about obesity were routinely illustrated by pictures of fat people in business suits, tucking into enormous steaks in upscale restaurants, this would portray obesity as a problem of overconsumption on the part of the rich, regardless of whether or not this was reflected in the content of the news stories themselves. In contrast, images of fat people in working clothes on a building site would be presenting a very different class identification of obesity.

Of the images on the BBC News website, just over half (55%) of the images of fat people and most (76%) of the images of thinner people had no class identifiers, usually because the image showed either rolls of naked fat or a thin naked midriff. This does not leave a very large number of images which could be given a class identification, but the pattern of these images was clear enough to be worthy of note. 45% of the anonymous images of fat people showed them clothed and with a context sufficient for an idea of their class theoretically to be conveyed. Only one of these showed anything which could be interpreted as a bourgeois or wealthy setting for the fat person concerned, and that was a cartoon. All the other images of fat people showed them as working-class, either through the clothes they were wearing or the location in which the picture was taken.

One of the most notable absences in the portrayal of the fat people in this survey was that of any images of fat people at work: there were no images of fat people in business suits, in offices or even in what might be viewed as more working-class workplaces like building sites or fast-food restaurants. The lack of fat bankers in particular is surprising, given the possibilities for visual representations of bankers' greed, but this only appeared once, in the lone cartoon. In contrast, the smaller number of images of thinner people did include the repeated appearance of an image of a woman in a distinctly fancy-looking gym, which was a markedly more middle-class image than any of the images of fat people.

The selection of the images of anonymous people illustrating these BBC News stories about obesity therefore appears to reflect a view of fat people as poor, or at the very least not wealthy, and unemployed. There are clearly intersections between these sorts of media portrayals of obesity and the reality for fat people, both in terms of the higher prevalence of obesity in poorer areas and the evidence of discrimination against fat people in employment, which does indeed raise the likelihood of any particular obese person being poor and unemployed. However, for the purposes of this discussion, it is sufficient to point out that in the view of media sources like the BBC, the fat and the working-class appear to be synonymous.

The portrayal of obesity as a working-class issue is not restricted to the media; it can also be seen in UK government reports on the issue. In January 2008, for example, the Cross-Government Obesity Unit issued a white paper called *Healthy Weight, Healthy Lives*.[37] The text of the report follows the common tack taken by official writing on obesity by discussing it primarily as a matter of individual responsibility but also as a consequence of 'modern lifestyles'. So, for example, the then-Health Secretary Alan Johnson wrote in the introduction to the report 'from the nature of the food that we eat, to the built environment through to the way our children live their lives, it is harder to avoid obesity in the modern environment'.[38] However, the pictures in the report send a different message.

This is a highly illustrated report, with thirteen full-page photos to only 37 pages of text. As might be expected from a government white paper, they are markedly aspirational: eleven of the images show people who are acceptably slim, and in five of these, they appear in the leafy settings of the countryside or parks. There are no images of extreme obesity of the headless fatty type discussed above, but there is one headless picture. It shows three young slightly plump black women from the neck down, standing in an urban street, holding carrier bags, eating fast food. It is one of the only three full-page images in the report in which the people shown could be seen as working-class, and forms the frontispiece for the chapter called 'The Challenge'.[39] The cumulative effect of the illustrations is to identify obesity with urban settings and particularly with poor, black people. Whatever the text might say about the generally obesegenic nature of society, the visual message about obesity here is that it is a class issue. In fact, the contrast between the urban setting for the 'challenge' and the rural or suburban setting for almost every other picture is so marked that, based on the pictures alone, you might think that government was arguing that the way to lose weight was to become middle-class and move to the suburbs.

The identification of obesity as a working-class problem is such that the UK government also uses obesity as a way of indicating class in other contexts. This could be seen, for example, in a Department of Work and Pensions (DWP) campaign from 2009 to encourage people to shop their neighbours for benefit fraud.[40] This campaign was clearly aimed at working-class communities, and the DWP went to some lengths to portray the 'benefit thieves' shown in their posters as working-class. The four different versions of the poster all showed their subject in what could be seen as a working-class setting: streets of terraced houses in the case of three of the images and a caff in the case of the fourth. However, in order to ram home the point, the subjects of the photos had other attributes which showed that they fitted into their environment: wearing a hoodie; leaning on a lamppost reading a tabloid; cleaning tables in the caff; or, in the case of the fourth poster, being fat. The portrayal of the fat woman in the fourth poster uses the assumption that obesity is a working-class issue to underline the class point of the ad campaign, and accompanies it with a subtext about fat people's problematic consumption stealing from the more morally restrained. Whatever the Obesity Unit might have said about the generality of the obesity, as far as the DWP is concerned, it is a class issue.

The reasons for the apparent correlation between fat and

poverty are, in the words of the Foresight report, 'poorly understood'.[41] Despite this, it would clearly be possible for the popular understanding of obesity as something which goes with poverty to lead to investigations into what it is about poor areas which might lead to a higher level of obesity among their residents, whether lack of access to decent, affordable food, lack of sports facilities or something else. However, this is not generally what is happening. The popular identification of fat and working-class does not appear to have given rise to a general view of obesity as a systemic issue rather than an individual problem. Rather, it remains predominantly a matter of individual responsibility, in which those who are fat can also be labelled as lazy and undisciplined.

These are, of course, the same terms in which working-class people are often described, whether the 'idle' unemployed, the DWP's 'benefit thieves', or anyone who raises their voice to complain about benefit cuts. As the women who agreed to go public against the benefit cap discovered, having children and needing benefits will get you labelled a 'scrounger' and lectured, with little evidence, on how you rely on the state but still have money for booze, fags and tattoos. This is not new; Marx pointed out in the nineteenth century how the capitalist class demanded that their workers practice industrious self-denial, in order to keep the level of wages they needed as low as possible.[42] It is however given new impetus in an age of austerity, when people who cost the government money are presented as the source of all economic ills.

It has been noted for some time that prejudices against fat people can serve as proxies for class or racial hatreds which aren't themselves expressed openly. Paul Campos examines this in *The Obesity Myth* with a discussion of an article about obesity which displays how the connections between fat and class can be used to justify prejudice against working-class people.[43] Greg Critser, the article's author, recounts a trip to a McDonald's in a 'lower middle income area of Pasadena... when the various urban caballeros, drywalleros and jardineros get off from work and head for a quick bite', and to a Krispy Kreme doughnut shop in what he describes as a Hispanic area in the San Fernando valley. As Campos points out, what is clear from Critser's accounts of his visits to both of these places is his disgust at the people eating in them, whether the 'stout Mayan donas' queuing at Krispy Kreme or 'the high risk population indulging in high risk behaviour' at McDonald's.

This view of fat, working-class people is not unusual, nor is the overt racial element to Critser's prejudice against the Hispanic people he describes. This article is only unusual because Critser goes on to express his class view of obesity openly. He comments that no one is fat in the company brochures produced by corporate America – evidence not of discrimination against fat people, apparently, but that fat people are too lazy to get decent jobs – and, in an extraordinary passage, even expresses bourgeois fear at the mass of the proletariat ranged against them: 'What do the fat, darker, exploited poor, with their unbridled primal appetites, have to offer us but a chance for we diet-and-shape-conscious folk to live vicariously?'

Critser's language is clearly extreme, but this view of the fat working class as a mass of unbridled appetites, damaging to the rest of us, is not, unfortunately, an outlier. It is clear from discussions of the effect of food production and consumption on the climate that the idea that the fat, problematic consumers are likely to be working-class is a common, if unstated, assumption. One demonstration of this is how often particularly problematic consumption is identified with junk or fast food, even when it doesn't necessarily follow. So, for example, a recent consideration of the environmental problems of overconsumption commented that the prices of consumer products like beef don't reflect the environmental consequences of their production. The identification of particular foods, meat especially, as key issues for food and climate change will be discussed in the next chapter, but note for the moment that while this statement doesn't make any explicit comment on the class of who is eating the beef, the idea that increasing the price would help solve the problem does seem to indicate that the problematic consumers may not have very much money. The comment goes on to link overconsumption of beef with 'overeating until obese' and says that the cheap price of beef explains why fast food chains can make so much money by supersizing their meals and is therefore why rates of obesity are rising worldwide.[44] Here, within the space of a paragraph, the argument has slipped from the price of beef, which you would think could be seen to encourage steak-eating in high-end restaurants as much as anything else, to fat people in McDonald's (where, incidentally, 'supersizing' a meal is as likely to increase the size of the portion of chips as the amount of beef), showing with whose consumption it is really concerned.

This attitude is expressed in a particularly forthright manner by Jonathan Porritt, in his introduction to Compassion in World Farming's 2004 report on overconsumption. Porritt's argument is that meat-eating should be strictly limited, as 'a moral outrage and a threat to ourselves, our planet's life-support systems and to future generations',[45] but what for him is especially egregious is not just meat *per se* but cheap meat: 'factor in all the health and food safety impacts of excessive meat consumption and the notion of cheap meat is revealed as the sick joke it really is'.[46] Despite the fact that he goes on to argue that 'we should all be eating a lot less meat', the concentration here on price suggests that what is problematic here is not bourgeois meat consumption, but consumption by the poor, who would not be able to afford to do it if it were priced 'properly'. This sentiment is repeated by Peter Melchett in the Soil Association's 2010 book on the food crisis, where he argues that a diet of 'huge quantities of cheap chicken, pork, dairy products and other mass-produced processed food...would saddle us with huge human, economic and environmental costs... [since] the crisis of diet-related diseases and obesity is already costing the NHS £7.7 billion per year', as if it was the cheapness of the food which determined its calorific content.[47]

In case anyone might have missed the point, Porritt strengthened the identification of problematic meat consumption as working-class meat consumption by the discussion of vegetarianism at the beginning of his introduction. Porritt, not at time of writing a vegetarian himself, expressed his struggles with his own meat consumption, which he could only overcome by sourcing all his meat as carefully as possible. Despite this, however, he was 'stuck in that tricky grey area between the moral elegance of vegetarianism on the one hand and the outright indifference of hamburger-guzzling omnivores on the other'.[48] The class identification of overconsumption could hardly be more clearly expressed.

What we are seeing here is the general panic about obesity becoming a vehicle for expressing ruling-class fears about the demands and appetites of the poor, fears which are particularly pointed at a time when working-class living standards are under attack. This does not mean, however, that the identification of the billion overweight people as the source of the problems of the food system is automatically incorrect. The idea that working-class people eating chips are responsible for starvation and climate change could be reality: the Western working class could really be eating the world to death. It would be an inconvenient truth, but is it the truth? The next step, obviously, is to examine the links between obesity and climate change.

Obesity as a cause of climate change

In some ways, a thematic link between climate change and obesity seems a fairly easy step. Given the common view of climate change as a disease of Western affluence, a product of our overconsumption and greed, it is obvious that fat people could appear as this idea's visual expression, the several embodiments of the perceived greed, sloth and irresponsibility of Western societies. From there it is only a small step to make fat people the physical representations of taking too much of the world's resources; robbing the starving billion of the food they need to survive; stepping too heavily upon the earth. Beyond this confluence of analogies, however, there have been some serious attempts to link obesity and ecological crisis.

Some of this has been at government level, where it seems mostly to arise from the tendency to express the seriousness of issues with reference to other issues more generally recognised as serious. An obvious example of this was when, in 2003, the US Surgeon General emphasised his view that obesity was the most pressing threat to health in the US by casting it as a worse threat than weapons of mass destruction.[49] He was not (as far as I know) arguing that there was a connection between fat and the war on terror, but trying to give the reporting journalists a soundbite they and their readers would understand. The connections made in official reports between obesity and climate change seem to be on the same level. In 2007, for example, the UK Government Office for Science issued the Foresight report on obesity, in which it called obesity 'the climate change of public health',[50] and in June 2009, 59 individuals and 39 organisations signed a letter to President Obama calling for a Presidential Commission on Healthy Weights, Healthy Lives, with the argument that 'the epidemic of overweight and obesity... is undermining the nation's health just as global warming is undermining the planet's health'.[51]

These are clearly analogies, rather than arguments for causal connections between the issues, although it also seems clear that the analogies arise from a perception that while the two issues may be separate, the sort of solutions which should be considered are similar. For the Foresight report, for example, while the measures it mentions, like improving public transport, are more focused on government policy than on individual action, the underlying connection between obesity and climate change is that it sees both as issues of individual behaviour, to be tackled primarily through individual choices: 'Obesity, like climate change, is a complex problem but it is not insoluble... tackling obesity is fundamentally an issue about healthy and sustainable living for current and future generations'.[52] The Foresight report does not mention dietary changes in examples of changes which could affect both issues, but the language of individual responsibility to adopt healthy and sustainable living is reflected in government reports on food and climate change. The Strategy Unit report of 2008, for example, echoes the Foresight report in its assertion that 'the environmental impacts of the food system are all, ultimately, a consequence of consumption decisions' and its argument for the adoption of 'a healthy, low-impact diet',[53] rather than any government-led systemic changes.

As far as the US and UK governments are concerned, then, obesity and climate change are unconnected issues thematically linked by their effect on lifestyle, rather than problems with any causal link. This is not the view of all researchers, as there have been some attempts to make a serious connection between fatness and the state of the climate. There is, remarkably, a subset of the literature on climate change and overconsumption which does attempt to argue that fat people are themselves, by their very fatness, a leading cause of climate change.

In May 2008, the BBC reported under the memorable title 'Obese blamed for world's ills' on a letter in *The Lancet* which argued that fat people's food and fuel consumption was driving up prices and production and therefore contributing to both rising food prices and climate change,[54] arguments which have been echoed in other recent articles.[55] The points made are various, but fall under two basic headings: that fat people engage disproportionately in climate-damaging behaviours, such as eating 'fattening' foods or driving large cars; and that their size itself means that they are personally responsible for high greenhouse gas emissions, regardless of their day-to-day choices, because higher weights mean more fuel use for transport, or because they require more calories to maintain their weight than others. Together, the message is that the possession of a 'healthy' BMI (Body Mass Index) is necessary for a low individual carbon footprint, and that anyone who lacks such a desirable possession should desist from the problematic behaviour in which it is assumed that fat people engage.

The problems with these arguments are numerous, and it is worth examining them in some detail. Michaelowa and Dransfeld open their article with figures for average daily per capita calorie consumption, which they state increased from 2,947kcal in the 1960s to 3,379kcal now.[56] While these are clearly presented as figures for food consumption, the authors don't explain how they were arrived at, and the 3,379kcal figure for current food consumption is similar enough to the 3,774/3,474kcal figure used by other overconsumption writers to suggest that they come from the same source. However, if this is the case, this is a figure for food production, specifically US food production, expressed as calories produced daily per head of population.[57] The assumption that the entirety of this food production is overconsumed by people eating themselves fat, (rather than, say, being cleared off supermarket shelves and thrown away), is just that, an assumption. The wide gap between even the figure for daily calorie consumption in the supposedly svelte 1960s – 2,947kcal – and the 2,000-2,500kcal recommended daily intake should also suggest that using the production figures as proxies for consumption may be problematic.

Michaelowa and Dransfeld themselves state, later in the same article, that 'there are insufficient data on the food intake of obese individuals' to assess the greenhouse emissions inherent in fat people's diets, and explain that they are in fact using figures for worldwide increases in consumption of high-fat foods as proxies for the supposed effects of fat people's diets on the climate.[58] The assumption that fat people eat more than others, or that an individual's consumption can be inferred from their body size, is frequently made. However, this conflation is not borne out by such evidence as does exist on the diets of obese people. A number of studies of obesity in children and adults have found that daily per capita calorie consumption appears to be declining; a 1997 study into the calorie intake of US adults, for example, found that the average per day was 1,785kcal, down from 1,854kcal per day. Since these studies are based on individuals reporting their own intake, there is of course the possibility of underreporting, that people don't realise, or don't admit, how much they are actually eating. This is a real problem for any studies based on individuals stating their food intake, but it has been pointed out that it cannot discount the clear decline in calorific intake found in several studies, in some cases since the 1930s. If anything, inaccuracies in individuals reporting their calorie intake would be expected to decline as people in general become more aware of the calorific values of different foods and more likely to keep track of them.[59]

It also seems that the declining average calorific intake is not simply a matter of thin people reducing their intake while fat people do not. A recent study of 1,400 eight-to-ten-year olds in a poor, largely Mexican-American area of the southwest US found that although 33% of the group were classified as obese, a staggering 44% of the children had daily calorie intakes which were so low that they would be classified as energy insufficient (less than 1,400kcal per day), while the average daily intake of 1,588kcal was also well short of the recommended daily average of 1,900kcal for children of this age.[60] In fact, studies have consistently failed to find any significant difference between the diets of obese and 'normal' weight subjects.[61] Received wisdom to the contrary, fat people on average don't eat any more than anyone else.

The argument that obesity is a problem for climate change because of the amount that obese people eat does not therefore appear sustainable. It is also worth noting the wide discrepancy between the figure for individual daily consumption, based on the amount of food produced, of 3,774kcal per day used in much of the writing on obesity within the overconsumption literature, and the 1,785kcal figure found by studies of individuals' diets. Even allowing for some underreporting in addition to the second figure, this suggests that waste may account for more of the difference between calories produced and calories needed in the US than is usually allowed. The gap between calories available and calories reportedly eaten is similar for the UK, where food production is 3,458kcal per head, but food consumption, based on self-reported data, is 1,807kcal.[62] While there will clearly be wide variations in the amounts and types of foods which individuals actually eat, the assumption that a significant number of individuals, fat or otherwise, are contributing to climate change by eating 3,774kcal per day does not seem one on which arguments about overconsumption and obesity can reasonably be based. The gap between food consumed and food produced highlights, of course, the wastage inherent in the system, indicating that the problem itself is more systemic than it is a matter of individual food choices. We will look at this again in more depth in chapter 5.

The argument that fat people's diets are a problem for climate change appears in fact to be a circular one. The diets of fat people are assumed to be high in greenhouse gas emissions, so it can then be concluded that, since they are high in greenhouse gas emissions, they are bad for the climate.[63] On examination, it appears that other arguments that fat people's behaviour is problematic for the environment are also based on similar circular assumptions. The claim, for example, that fat people are responsible for higher transport emissions than thinner people is made, it transpires, with the help of highly dubious assumptions about behaviour: 'we assumed that all individuals with a BMI <30 use an average small car (e.g. Ford Fiesta) and that individuals with BMI >30 use a car with more internal space (e.g. Ford Galaxy)'.[64] The same study also assumed that fatter people would walk less and drive more, because 'walking is an effort for heavier people'.[65]

These assumptions enable the authors of the study to ascribe the higher emissions from the larger cars to fat people, but without any evidence to justify this. The notion that fat people are more likely to drive than not seems to be relatively common: Charlotte Cooper recollected how, at one of the first fat studies conferences she attended, a (thin) woman in the audience pronounced with unshakeable authority that 'fat people don't use public transport'. I have not, however, been able to locate any study establishing this fact, which would of course fly in the face of other stereotypes about fat people, such as that they are more likely to be working-class and therefore more likely to use modes of transport on which the poor disproportionately rely.

Insofar as they are based on any logical reasoning, these notions seem to contain an idea of people with BMIs over thirty as so fat as to be practically incapacitated; an entirely unrealistic idea which applies a degree of fatness experienced only by a very small percentage of individuals to all those whose BMIs categorise them as 'obese'.[66] The idea that 'carrying' fat around is just like carrying heavy weights is a familiar one in obesity studies, and there have been studies which have attempted to model the energy costs of walking for fat people by putting thin people into fat suits.[67] However, it is possible that larger people may walk in such a way as to minimise their overall energy costs, and may therefore not use up as much energy walking as a thin person who believes that by lugging a heavy backpack they are simulating fat people's experience.[68]

The evidence for obesity as a genuine problem for climate change appears to be slight, as would perhaps be accepted by the authors of the studies quoted here. Edwards and Roberts conclude their article by admitting that 'we have made a number of assumptions, all of which can be questioned'.[69] It is also worth noting that the actions suggested in these articles to deal with the problem tend not to address the supposedly problematic behaviours of the obese directly. Michaelowa and Dransfeld's advice, for example, is not for obese individuals to abjure the fattening foods which cause their disproportionate effect on climate change, but to adopt 'one hour of daily cycling or walking'[70]; a worthwhile programme, perhaps, but not one which would affect the majority of the problems outlined in the article.

That these arguments are not particularly convincing in their own terms may not have escaped some of the authors themselves. Michaelowa and Dransfeld for example conclude their article with the justification that, although people are resistant to being told by governments to lose weight, being told that they are also destroying the planet with their fatness might be the spur they need to get on a diet: the link between the issues, in other words, is that they can be combined to exert additional moral pressure on those deemed to be at fault, not actually because there is a causal connection between obesity and climate change.[71]

In the light of this degree of uncertainty from these studies' own authors about their conclusions, it is reasonable to wonder how it is that this little sub-field of studies of overconsumption and its effects on ecology arose in the first place. How did it ever seem reasonable to posit that the ecological crisis facing the world might be the responsibility of people who weigh more than average? How is it that we have arrived at a position where major arguments about the causes of climate change see it both as a problem caused by affluence and as the responsibility of the poorest people in the Western world? To understand how we have got here, it is necessary to review how accounts of the problems of the food system, for the climate and for feeding the world have developed.

2

How the problem with food changed

In 2006, the UN Food and Agriculture Organisation (FAO) issued a report called *Livestock's Long Shadow* about animal farming's effect on the environment.[72] Its message was a stark one. Animal farming is causing serious harm to the environment, polluting water supplies, driving deforestation and desertification and causing 18% of global greenhouse gas emissions, including 30-40% of the methane, one of the most potent greenhouse gases. It is not only contributing to climate change but also indirectly to hunger. The FAO did not think that animal feed is taking food directly from poor people's mouths, but it nevertheless concluded that livestock eat more human edible food than they produce, and this competition for grain can only increase prices.[73]

The problem seemed particularly acute because meat consumption is on the rise. There is, the report concluded, a global trend towards dietary convergence: 'similar eating habits, such as fast and convenience food are catching hold almost everywhere'.[74] This means that meat-eating per head in the developing world has doubled since 1980, while the total meat supply has tripled.[75] In the view of the FAO, the luxury tastes of a more affluent world population were causing a serious problem which it was urgent to address. The conclusion of the report was as outspoken as it is possible for an official UN report to be:

> Given the planet's finite natural resources, and the additional demands on the environment from a growing and wealthier world population, it is imperative for the livestock sector to move rapidly towards far reaching change.[76]

Since the FAO report came out, their conclusions have been reiterated and amplified in a number of other works. These have generally widened the focus from livestock farming to food production in general. So, in 2008, Greenpeace concluded that agriculture's contribution to greenhouse gas emissions, including direct and indirect emissions, was somewhere between 17% and 32%;[77] the Strategy Unit attributed an 18% share of greenhouse gas emissions to 'the food chain';[78] the Food Climate Research Network concluded that the food system was responsible for 19% of UK greenhouse gas emissions;[79] while Zero Carbon Britain went with the 18% figure for global agriculture emissions.[80]

These figures put food production relatively high on the list of major contributors to climate change. The FAO argued that livestock's contribution to greenhouse gas emissions was greater than that of transport,[81] a conclusion repeated in much of the coverage of the report, with headlines like 'Cow emissions more damaging to planet than CO^2 from cars'[82] or the pithier 'Eating meat is worse than driving a truck... for the climate'.[83] Other calculations still put transport slightly ahead (or should that be behind?) – George Monbiot, for example, put UK land transport emissions at 22% of the total[84] – but in either case, food's greenhouse gas emissions are clearly significant.

If food is a climate villain, some foods are more villainous than others. Despite the shift from livestock production in the FAO report to food production more generally in later works, meat remains a particularly problematic foodstuff for all the reports. The Food Climate Action Network report, for example, cites a study of the Dutch food system which found that meat was responsible for 28% and dairy for 23% of food-related emissions.[85] These emissions have a variety of sources, but the overall conclusion is that cows are a problem for the climate. Deforestation to free up land for cattle grazing, or to grow cattle feed, contributes to climate change; cattle feed is grown using fertiliser, the production of which emits a potent greenhouse gas, nitrous oxide; and worst of all, cows fart methane. The 30-40% of methane emissions which can be attributed to the livestock sector largely come from the digestive tracts of ruminants.

Crucially, this identification of the existence of cows as problematic places the principal issue at a particular point in the food chain. The Food Climate Action Network report made this explicit: most emissions, according to this report, come from agriculture, the initial stage in the chain which goes from the field to the supermarket shelf. This accounts for more than transporting, processing, packaging and selling the food put together, and should therefore be seen as by the far the most important aspect of food production.[86]

From food miles to cows

What this means is that it is the essential characteristics of certain types of food that are problematic, not the way they are treated within the modern food system. Meat and dairy products cause emissions by their very nature, not largely because of their production methods. Recent reports differ on the extent to which these emissions could be reduced, other than by eliminating the offending animals. Greenpeace's report, for example, took a reasonably favourable view of the possibility of mitigating emissions through changing farming practices, on which the Food Climate Action Network report was less keen.[87] However, all of these arguments see the problem with meat as arising from the nature of meat whatever production methods are used, rather than, say, from issues arising from globalisation and industrial food production.

This marks a fundamental shift from earlier thinking about food and the environment. Previous arguments about food centred on the environmental cost of industrialised food production and distribution within a global food market.[88] In this view, there is nothing existentially wrong with types of food; the problem is the introduction of an unsustainable energy source – fossil fuels – into a food system which by its nature is essentially sustainable. So, for example, a 2000 report on trade, agriculture and climate change concluded that the modernisation of food production had tied what should be a renewable source – farmland – to non-renewable oil supplies, and made it a significant contributor to climate change in the process.[89] Similarly, in 2001, Sustain commented in their report on food production that the unsustainable nature of modern food supply arose from its reliance on oil: 'Reliance on an energy source that is consumed more quickly that it can be regenerated is obviously not sustainable. The present system can only exist as long as inexpensive fossil fuels are available'.[90] The global market has turned food production into an enormously profitable and enormously damaging industry, but in this view, there is nothing intrinsically wrong with food production *per se*.

Fossil fuels appear at all stages of the industrialised food chain: modern international farming operations, Sustain observed, are 'often large-scale and based on industrial techniques, requiring high levels of inputs such as feed, pesticides, fertilisers and machinery'.[91] However, in this view, the key is the transport of food, and the packaging and production generated by a global supply chain which transports foods for long distances. Sustain highlighted the growth of the international food trade, which trebled between 1970 and 2000.[92] Similarly, in the same year, the Green Party described 'the great food swap', where, encouraged by the Common Agricultural Policy, European countries exported and imported often remarkably similar amounts of the same foods to and from each other. The report gives examples like the UK poultry trade with the Netherlands, where in 2000 Britain imported 61,400 tonnes of poultry meat and exported 33,100. Overall British pork and lamb exports showed the same pattern, with 195,000 tonnes of pork and 102,000 of lamb going out, while 240,000 and 125,000 tonnes came back in.[93]

The problem was not restricted to import and export, as production of, particularly, processed foodstuffs was increasingly globalised, so as to involve transport at many intermediate stages as well as from farm gate and to the shop. Sustain gave the example of a Swedish tomato ketchup: it was made from tomatoes grown in Italy (production which was itself sustainable and relatively free from fossil fuel use and pollution), which were then packaged in bags made in the Netherlands, driven to Sweden to be bottled in bottles made in the UK and Sweden from materials made in Japan, Italy, Belgium, the USA and Denmark, then capped with caps which were made in Denmark and transported to Sweden by road freight.[94] While the arguments have moved on, this sort of production and distribution has clearly not: Felicity Lawrence, for example, found some later examples of similar practices, like the chives in an M&S vegetable dish which were being flown to Kenya to be tied around vegetable bundles, and then flown back to the UK.[95]

The solution to this was seen as the creation of a localised, as opposed to globalised, food system. Reducing the distance food travelled would not only reduce the greenhouse gas emissions, and reliance on fossil fuels, intrinsic to long distance freight, but would also enable the creation of organic, sustainable farming practices, for which small-scale, diverse farming was key. Sustain's recommendations for action, for example, covered actions by consumers, farmers and governments, but for all of these focused on reducing food miles.[96] By emphasising the environmental damage caused not by transporting meat and dairy products but by producing them, the recent works on food and climate change are clearly calling for different priorities. Not only that, but the argument that the most problematic aspect of food production is the existence of cows, not transport or packaging, is being made with an explicit break from the past prioritisation of food miles issues.

As noted above, there has been a tendency for the new importance of food for climate change studies to be expressed in terms of transport. In 2006, for example, an article in *Earth Interactions* assessed the difference in greenhouse gas emissions attributable to a vegan and a red meat diet as being equivalent to the difference in emissions from a sedan and an SUV,[97] while *The Independent* reported that food had overtaken transport in the greenhouse gas emissions league table, to be 'more than cars, planes and all other forms of transport put together'.[98] In 2012, *The Independent* was at it again, covering the launch of a new report on the greenhouse gas emissions associated with different foods with the claim that if the entire British population went vegetarian or vegan, this would be the equivalent of taking half the cars off the roads.[99] This is presumably at least partly an example of the journalistic trick of explaining a new concept in terms of something more familiar. Just as sizes of oil slicks are expressed in terms of the size of Wales, or dinosaurs in double-decker buses,[100] it may be that transport's contribution to climate change is so well understood that it can be used as a measure of everything else's emissions. However, there is also a more deliberate comparison going on, one which serves to distance the new arguments about food from those about food miles.

The Food Climate Action Network are up front about how far their work entails a departure from the food miles agenda, commenting that 'the more appropriate focus of concern might not be how far our food has travelled but the proportions of different foods on our plate'.[101] Even when their report considers the issues of processing and packaging which are so important in reports like Sustain's, the Food Climate Action Network sees them in terms of the nature of the foods being treated, not as processes with which food production could dispense. Refrigeration, for example, for Sustain is used more extensively in the modern food system because it is needed to keep food fresh while it is transported across the world. For the Food Climate Action Network, the question is still about the essential nature of some foods: 'what is it about the foods we eat and the way we manage our lives that renders refrigeration necessary?'[102] This approach can also be seen in other works on food and climate change in the last few years. A 2008 article in *Environmental Science and Technology* compared the effects of avoiding food miles or avoiding meat, and concluded that transporting food contributed only 11% of the food chain's greenhouse gas emissions and that the 'average US household' could achieve the same effect as entirely eliminating food miles simply by cutting out one day's meat consumption.[103] In this context, the titling of a 2012 report into the effects of a mass shift to a vegetarian or vegan diet as 'the impacts of realistic dietary choices' is interesting. The article does not make the comparison between the greenhouse gases produced from food types and food miles; indeed, its comparison between the effects of food and transport is explicitly limited to private cars. What it does, however, is couch all the diets considered in terms of what could be bought solely from supermarkets, and includes the assumption that they will include air-freighted and polytunnel produce. By the end of the article, the 'realistic' of the title appears to refer as much to the impossibility of changing the food production system, which was the target of the food miles campaign, as to individual food choices.[104] Thus the break with thinking about food's contribution to climate change in terms of food miles is clearly also a shift from thinking about it in terms of the food system to an issue of individual lifestyle choices.

Globalisation out of the spotlight

One obvious consequence of this is that different types of food are put in the spotlight. In terms of globalisation and fossil fuel energy use, the most problematic food groups are food like cereals, chocolate and coffee, which require high levels of processing, and therefore take a lot of energy to produce. In 2000, it was estimated that breakfast cereals took 15,675 calories (kcal) to produce one kilogram of finished product, with chocolate taking 18,591kcal and coffee 18,948. Beef does not come out of these calculations particularly well, requiring 35 times more calories to produce than it provides, but overall one kilogram of meat only required 1,206kcal, less than frozen fish, baked goods, frozen fruit and vegetables and dried processed foods.[105] In contrast, the argument that meat and dairy products are essentially problematic for the climate puts comparatively little emphasis on processed and highly packaged foods, as well as explicitly minimising the importance of food transportation. Locally-sourced beef, it seems, is a worse choice from a climate point of view than air-freighted strawberries.

More importantly, the different assessments of the most serious problems with food also imply different sorts of actions to solve them. If the issue is how food is traded and distributed, the solutions most obviously lie at the level at which trade policies are developed, i.e. government. It is true that the food miles arguments did give rise to calls for people to change their individual behaviour, by buying locally-sourced, in-season produce, preferably from farmers' markets rather than from supermarkets. This is all in the Sustain report, but while individual consumers could 'help increase real choice' by supporting local food businesses, the longest list of recommendations for action is for governments to 'take sustainable development seriously': this list contains sixteen items compared to eight for individuals, three for farmers and four each for retailers and local authorities.[106]

It is not simply that governments were seen as best placed to address the problems caused by the food system, but that the food miles issue was part of wider concerns about globalisation. That the problems of global food supply are not inherent but are a function of the market is a central point of the food miles arguments, and an intensely political one. So it is that Sustain cite the increasing importation of fresh vegetables from Kenya, which leapt by 30% between 1979 and 1999, not just in terms of the environmental effects of the transport and packaging, but also in terms of the poverty in Kenya caused by the consequent distortion of Kenyan farming. They point out that during this period, Kenyan vegetable production was centralised in the hands of just three producers, a development which was supported by UK supermarkets, and which had the effect of further impoverishing already poor farmers and driving people off the land. By 1999, consumption of vegetables actually in Kenya had gone down by 39%, because people could no longer afford to consume what they were producing for export. The report concludes: 'where hunger exists, what is often lacking is not food but access to it – either having the money to buy it or the land to grow it on.'[107]

The development of this global food market was of course not an isolated one, but part of the globalisation of the world economy over the last thirty years of the twentieth century. The food system criticised in reports like Sustain's was explicitly advanced by international agreements like the Agreement on Agriculture within the General Agreement on Tariffs and Trade (GATT), one of the main weapons in the World Trade Organisation (WTO)'s drive for full trade liberalisation and globalisation. The Agreement on Agriculture aimed to create and consolidate a global agricultural economy. As Colin Hines put it in his call for localisation, in this system 'all countries produce specialised agricultural commodities, and supply their food needs by shopping in the global marketplace. Food is grown, not by farmers for local consumers, but by corporations for global markets.'[108] Calls for a shift to local food production can therefore be seen as part of a wider anti-globalisation agenda, whether or not this connection is made explicitly.

This means that discussions of the problems of food in terms of food miles need to be read against the background not only of the wider criticisms of globalisation, but also of the anti-capitalist movement which came onto the streets to protest at the WTO meeting in Seattle for five days from 30th November 1999.[109] As Felicity Lawrence noted in 2004, 'many of the ills of the current food system are ones foreseen by the anti-globalists and the anti-capitalists'[110] and environmental issues were a central concern of the anti-capitalist movement from the start. Jeffrey St Clair remembered in his Seattle diary how the 'direct action warriors on the front lines' included people from Earth First, the Alliance for Sustainable Jobs and the Environment and from Rainforest Action Network, while the 'robust international contingent' on the streets was made up of 'French farmers, Korean greens, Canadian wheat growers, Mexican environmentalists, Chinese dissidents, Ecuadorian anti-dam organizers, U'wa tribespeople from the Colombian rainforest and British campaigners against genetically-modified foods'.[111] The French farmers were led by José Bové, who famously attacked a McDonald's under construction in Larzac, France in 1999 in protest at the US imposition of heavy import duties on Roquefort cheese. Jeffrey St Clair recounts how in Seattle, Bové handed out rounds of Roquefort to the protestors outside the Seattle McDonald's, made a speech against GM foods and watched as the crowd then broke the McDonald's windows and urged diners and staff to join the protests.[112]

It is notable from the list of international campaigners that it was not just green issues but issues of food and its effects on the environment which were high on the Seattle agenda. Food miles in fact have a significant place in anti-globalisation analyses. The environmental costs of food transport are not simply one more problem of globalisation but an important indication of how orthodox economics was unable to account for the environmental damage the 'privatisation, commodification and rationalisation of the globe' was causing.[113] Localisation agendas are not always particularly progressive – calling for a return to local food production can come after all from a backward-looking, romantic sensibility – but in the anti-capitalist movement there is a clear link between localisation of food as an alternative to food miles and an analysis of the entire globalising system as flawed.

In a context in which concern about supermarket transport is part of the issue mobilising hundreds of thousands on the streets of Seattle, Washington DC, Genoa (where 300,000 people besieged delegates to the G8 in the centre of the city for three days) and other cities, the obvious actions for people incensed by the damage being caused by the food system, or inspired by the calls for local food production, became two-fold. They would include changing their own food choices, but would also naturally raise the question of involvement in this wider movement. The arguments about food miles were political arguments, part of a movement fighting for a complete system change.

This political nature is apparent even in the measured prose of the Sustain report. In their discussion of the UK government's actions on food, they commented that more trade liberalisation would mean more large farms, industrialised production and transportation, the precise opposite of the required transition to sustainable production, and that the government was not helping to introduce sustainability. The passage essentially implies that the government is in hock to agribusiness, suggesting that the long list of actions needed from government would not be achieved without pressure from the anti-capitalist movement.[114]

Meat and individual choice

In contrast, if the problem with food and climate change arises from the essential nature of meat and dairy products, there is no point railing against the global food system. When it comes down to it, cows will fart methane whatever economic system they are kept in. The target of the recent arguments about food and climate change therefore becomes not the food system but individual meat-eating. It's notable that cutting meat and dairy consumption is the key 'high priority' action listed by the Food Climate Action Network report,[115] and is also prominent in the Greenpeace report.[116]

It may seem obvious that the main recommended solution to a problem caused by meat production would be for individuals to be persuaded to choose to eat less meat, but this isn't the only way in which this could be addressed. A shift to a low-meat diet by much of the population in the UK would require a significant change in food supply and distribution. Food deserts – poor areas in which decent, fresh food is virtually unavailable – are a reality, and for many people, the cost of fresh, nutritious food can be prohibitive compared to 'junk' food. Calls for less meat-eating for the sake of the climate could address these and other such issues, recognising that people don't choose what to eat in a vacuum, but as part of a food system in which their choices are shaped by forces beyond the individual at the point of purchase. The fact that the food and climate issue is presented largely as an issue of people's individual choice to eat meat, and not as an issue requiring, for example, campaigns for better access to decent food, is demonstrative of how this argument marks a shift in thinking from the political food-miles campaigns to a more individual-based approach.

The aim, in fact, is sometimes explicitly presented as being action by the few, not the many. In 2004, for example, Compassion in World Farming issued a report calling for cuts in meat and dairy consumption, but without imagining that this would be widely adopted. Instead, those who were enlightened enough to adopt a vegetarian or vegan diet would do the work for those who had not already seen the light: 'One person's 100 percent reduction can help to "subsidise" six people who haven't yet reduced their meat consumption at all'.[117]

This matters particularly because the effect of a shift to a more individual-consumption approach to the problem of food is not necessarily restricted to purely this issue. The popularisation of the food arguments tends to present reducing meat consumption as a potential focus of individual activity on climate change in general. Promoting the 'Meatless Mondays' campaign in June 2009, for example, Paul McCartney argued 'many of us feel helpless in the face of environmental challenges, and it can be hard to know... what we can do to make a meaningful contribution to a cleaner, more sustainable, healthier world. Having one designated meat-free day a week is a meaningful change that everyone can make.'[118] The Soil Association repeated this in their call for 'Positive Action to Prevent a Global Food Crisis' in 2010: 'Issues like climate change and peak oil can sometimes make us feel powerless; but food is different... we can all do something to change our relationship with food'.[119] This is in effect another strand of argument for why food should be a major focus of efforts to deal with climate change – not only is food production a significant cause of emissions, but it provides an easy way for everyone to get involved. In so doing, however, it imposes a particular approach to climate issues based on individual rather than collective action.

The 'change that everyone can make'?

One effect of the 'change that everyone can make' approach to food and climate change has been the simplification of complex arguments about the climate effects of different sorts of meat production into a simple 'eat less meat' message. This is often accompanied by an elision of the significant difference in these terms between a vegetarian and a vegan diet. In July 2010, for example, Annette Pinner, chief executive of the Vegetarian Society, told the *New Scientist* that 'the most effective way to reduce the environmental impact of diet, on a personal level, is to become vegetarian or vegan'.[120] This is despite the fact that the dairy products included in the vegetarian diet would largely come from cows and therefore share much of the climate contributions of beef. In order to support a vegetarian but not a vegan population, we would have to keep the cows.[121]

Cows, we know, are the worst climate criminals because of their methane emissions. Other meat animals are responsible for lower emissions: sheep also fart methane, but are mostly still fed on grass, so sheep farming does not have the issues of deforestation and diversion of human food crops that come with cattle feed. Pigs and chickens don't emit methane, although there are issues with disposing of the waste from intensive pig farms, and they are also more efficient than cows in terms of feed. Ruminants are not the only food-related source of methane, as rice cultivation is also a significant source, although rarely mentioned in these arguments.[122] A dietary shift from beef to pork and chicken would reduce greenhouse gas emissions,[123] but since cows are also the main source of dairy products, a shift from meat to a vegetarian (as opposed to vegan) diet would not. In 2006, two US academics, Gideon Eshel and Pamela Martin, compared the greenhouse gas implications of different styles of diet, and found that a lacto-ovo vegetarian diet would have higher CO^2 equivalent emissions per calorie than a diet with the same calorific total which excluded dairy but included poultry or fish.[124] This understanding is built into the recommendations of the Zero Carbon Britain report, which calls for a 90% reduction in UK beef cattle and 80% reductions each in dairy cattle and sheep, but which allows 100% of the current levels of laying poultry, 50% of table poultry and 80% of pigs. In so doing, however, the report also demonstrates how deeply engrained the idea that meat generally is problematic has become. It is difficult to think of another reason for reductions in chickens for meat, while egg-laying chickens are exempt.[125]

It's worth noting in light of this that a large part of meat consumption worldwide is actually pork or chicken consumption. Chicken in particular is the growth meat in both the developed and the developing world. Chicken consumption in the UK has increased by 20% in the last two decades, a period in which the absolute consumption of other meats has remained steady despite an increase in population,[126] while the largest single share of the 200% increase in US meat consumption between 1950 and 2000 was also attributable to chicken.[127] In 2002, the UN FAO estimated that 77% of the increase in meat production in developing countries was from chicken and pork.[128] Even dividing the meat category into red and white meat doesn't mean that all red meat consumption is the most problematic kind: 32% of the 'average US' diet is made up of red meat, but this consists of 61% pork, and only 38% ruminants (2% lamb and 36% beef).[129]

The variance in emissions levels between the different kinds of meat aren't of course unknown to the various meat-free campaigns, some of which point this out explicitly. The 10:10 campaign, for example, doesn't go so far as to differentiate between beef and chicken in its recommendation to have one meat-free day a week, but does advise 'don't replace with just-as-bad cheese'.[130] There is plenty of literature arguing that, in trying to persuade people to change their behaviour to deal with climate change, a simple, clear message is best,[131] and presumably this is why popularisations of the climate effects of beef have become messages about all meat, only patchily aware of the issue of cheese. It's evident that giving up meat for one day a week is supposed to lead at least some participants on to further cuts in meat consumption, and it may well be assumed that once some people have adopted vegetarianism they will then be motivated to adopt a more vegan diet as well. In 2004, for example, Compassion in World Farming recommended that everyone reduce their meat consumption by at least 15%, but move on to further cuts, including reducing their milk consumption by 25%.[132] This underlines, however, how far this is from a collective approach, as only the most committed and enlightened would presumably go on to make extensive changes to their diets.

Turning the arguments about meat and climate change into a punchy, persuasive message means that message has a rather distant relationship at times with the actuality of emissions from different types of meat production. This is not the only issue with understanding the problem of food and climate change in terms of individual food choices. The idea that problems caused by meat production could be addressed by individual decisions to eat less of it is based on an assumption that production is led by consumer demand: we decide what we want to buy, and the market provides it. Where the way to achieve reductions in meat consumption is addressed specifically, it tends to be market mechanisms which are invoked: Zero Carbon Britain, for example, comments that carbon-pricing would do the job, as 'consumers too would be reorienting their food choices in response to unmistakable price signals'.[133] It is not said here, but it would of course be working-class people who would be the most vulnerable to these price signals; another reminder of which class of individual choices we are actually talking about.

Supply and demand

Arguments that the market is all-powerful have taken a bit of a hit since the collapse of Lehman Brothers in September 2008 ushered in the economic crisis. As far as food is concerned, it is not the case that people simply decide how they want to eat and the market rushes to supply it. Production can lead consumption, and the consumer we're presented with by the neo-liberal model, who makes consumption decisions in the free market based on rational self-interest, does not bear much relation to reality. This issue will be pursued further in the next chapter, but it is worth noting here some inconsistencies in the arguments about food consumption patterns and climate change.

Underlying much of the recent literature on food and climate change is the powerful assumption that the 'Western' diet – lots of red meat, dairy, fat, sugar and processed foods – is inherently the most desirable diet in the short term. Even though as usually characterised it is hardly the healthiest way for most people to eat, it seems to be viewed as the one which delivers the most immediate gratification. The assumption is that this type of diet is the one to which everyone in the world would gravitate if they were able to do so. You would think that examples of different diets only exist because poverty or lack of Westernisation have protected people from the most desirable model, or, if we are talking about residents of Western countries or affluent classes in the developing world, because they exercise virtuous self-restraint.

One demonstration of this assumption is the familiar claim that a growing middle class in countries like India and China will drive environmental destruction as they become wealthier. Their increasing wealth will lead them to adopt the 'Western' diet; a diet which is so resource-heavy that if everyone were to eat this way, we would need four planets to supply us.[134] To be clear, I am not suggesting that consumption of meat generally, or fast food specifically, cannot be shown to be increasing among relatively prosperous Indians and Chinese, or indeed elsewhere in the developing world, although it is worth noting that demand for meat and dairy in India and China overall seems to have increased less rapidly in 2002-08 than it did in 1995-2001.[135] However, shifts to a more Western diet are usually portrayed as simply a result of that relative prosperity, assuming both that this diet is automatically more desirable than traditional ways of eating and that food production follows the desires of these groups to eat like Westerners. This is not a safe assumption.

Japan, for example, is one country which saw a marked increase in meat consumption in the 1970s and 1980s, and where there is a market for Western fast food companies like McDonald's. However, this dietary shift was part of US influence on the Japanese economy and society after the Second World War, and was at least partly driven by both Western corporations wishing to expand their markets in Japan and by Japanese reformers who associated Westernness with national pride and success. It was argued by some Japanese, for example, in the late 1960s that the Japanese would never become physically strong enough to compete with meat-eating Westerners if they did not eat meat themselves.[136] This was picked up in the marketing of fast food and other 'Western' products: when McDonald's opened a restaurant in the Tokyo Milsukoshi department store in 1971, the slogan promised diners that 'If you keep eating hamburgers, you will become blond'.[137]

The shift from the Japanese diet clearly cannot be reduced to the inherent desirability of hamburgers, nor to the newfound ability of 1970s Japanese to afford them, stimulating the market to provide. While the details of the spread of US food in Japan is clearly specific to that country, there is a more general connection between the spread of multinational food brands and US imperialist interests. While it is assumed that consumers around the world adopt diets full of junk food because humans are naturally predisposed to eat that way if they can, looked at from the other direction, it is clear that shifts to a more Western-style diet are driven by the corporations who will profit from them and by foreign policy concerns. US foods have been marketed at populations as part of attempts to inculcate pro-US, pro-Western attitudes in countries in which the US had foreign policy interests.[138] When Thomas Friedman famously asserted in 1999 that 'No two countries that both had McDonald's had fought a war against each other since each got its McDonald's',[139] despite his claim that this represented merely the effects of growing middle classes across the world, it was clearly an acknowledgment of the role McDonald's played in US foreign policy. As he remarked later in the same book, 'in most societies, people cannot distinguish any more between American power, American exports, American cultural assaults, American cultural exports and plain vanilla globalization. They are now all wrapped into one.'[140]

In the same way, the 'Western' diet itself is not the product of the desires of its consumers unmediated by food producers, corporations or any other interests. It is clear, in fact, that many of the salient features of this diet are creations of corporate interests rather than people's greed or self-indulgence. The relatively high beef consumption in the 'typical US diet', for example, is not an accidental development, but marks a shift from pork-eating encouraged by beef and grain companies. Beef producers were attracted to the mid-West plains in the US when they were left free for cattle by the extinction of the buffalo. When the grassland contracted at the end of the nineteenth century, the cattle were fed largely on grain, and grain and beef production became so integrated that they were often carried out by the same companies. Cattle feed remained such a profitable destination for grain production that from the late 1950s there was a concerted effort to increase beef consumption, often through supermarkets which were themselves owned by beef production companies. Subsequently, beef producers, specifically hamburger producers, were able to use beef fed on Latin American grasslands to undercut grain-fed beef and beef's main competitor, pork. Beef-eating was so profitable that it was in companies' interests to ensure that US consumption remained high. Consumers may have felt that they were making a free choice to eat hamburgers, but there was in fact a concerted effort to encourage them to do so.[141]

The use of Western food corporations to further Western interests in other parts of the world makes it evident that governments understand how populations' diets develop as part of a complex system, of which individuals' rational and uninfluenced decisions on what they want to eat are only a small part. However, when discussing food, the UK government at least takes the view that it is consumer decisions which drive everything else about food production. In 2008, for example, the Strategy Unit reported in *Food Matters* on the effect of the food chain on greenhouse gas emissions and health issues around food. Their belief in the centrality of consumer behaviour in all these issues was made clear from the first page, as the authors complained that consumers 'expect retailers, manufacturers or the Government to act on their behalf and to "edit" problems out of the system rather than ask them to choose'.[142] This was a swipe at food-miles-type criticisms of the food system as a global system, and the Strategy Unit was very clear that this approach is misguided. There is no alternative to the free market for food – the UK government's response to the food crises of 2007-08 was to call for more free trade internationally in agricultural products and food – so the government cannot control or change the system. The only role for government, apparently, is to 'support consumers in the choices they make',[143] although they did accept a role in unspecified action to 'avoid market failures'.[144]

This neatly encapsulates both the view that food's contribution to climate change is caused by individuals' choices to eat some types of food rather than others, and the claim that those choices drive a free market in global food production into which governments cannot intervene. The report went on to say explicitly that 'the environmental impacts of the food system are all, ultimately, a consequence of consumption decisions' – food production is created by consumption decisions, which are therefore ultimately responsible for the effects of the production.[145] The vision of the ideal future food society therefore hinges on consumers being more informed and making 'better' consumption decisions, as set out in this description of this neo-liberal utopia:

> It will be a system with high levels of trust, challenged and supported by civil society but with the means and the will to work through problems through informed debate rather than conflict and confusion.[146]

Clearly, there will be no need for antagonistic, confusion-creating campaigns about food miles.

The thinking set out in the Strategy Unit report is typical of government approaches to climate change in general. Other climate-change-related campaigns, like Act On CO^2, have taken a similar view that greenhouse gas emissions can be reduced to individual behaviour. The Act On CO^2 website focused on small changes to individual lifestyles; their TV ad campaign in late 2009 to early 2010, for example, was exhorting us to 'drive 5 miles less a week', backed up on their website by 'helpful' tips on how to achieve this, like 'plan your journey in advance' and car-sharing.[147] This was justified by the statement that 'around 40% of CO^2 emissions in the UK are caused by things we do as individuals'. It can be supposed, however, that this approach reflects not so much an assessment of the statistics on greenhouse gas emissions, but a governmental desire to push responsibility for action on climate change on to individuals and away from itself.

In promoting individual actions to the exclusion of all others, the government was not only ignoring how social systems determine those supposedly free consumer choices, but espousing a neo-liberal view of the state as nothing more than the facilitator of the market. This is clearly a very different approach from that taken by the food-miles-type arguments, which is unsurprising considering the extent to which the latter were part of the general anti-globalisation arguments of the anti-capitalist movement. However, the arguments that the problem of food and climate change arises from the production of specific types of food, rather than from the global food market, support the government's conclusion that it's all down to consumers. This is a significant shift in the political meaning of the food and climate change issue. With the food miles arguments, modern food production's contribution to climate change was another reason for fighting the system. Now, the conclusions of reports like the Food Climate Action Network's are remarkably similar to those produced by government, as they seem united in an approach which places consumer decisions in the free market at the centre of food and climate change issues.

This is not only apparent in the shift from a view of the problem as systemic to a matter of individual consumption decisions. As noted above, the change in the focus of the food arguments has also entailed a shift in the types of food seen as particularly problematic, from highly processed foods like chocolate and coffee to meat and dairy products. This is significant as it enables the refutation of not only the idea that food miles are harmful, but also that modern food-processing could be harmful, and that a switch to less processed, less ready-made food would be preferable. Alongside the food miles arguments is therefore an assumption that small scale, organic production of fresh food for people to prepare themselves at home is better than large-scale production for supermarket ready meals. Supermarkets in these earlier food arguments appear as the source of many of the food miles: up to 40% of lorries travelling in the UK in 2004 were involved in food production and distribution, and many of them would have been from supermarkets, many of which had made substantial increases to their fleets in the previous few years.[148] This was however not their only contribution to food and environmental problems, as they were also seen as preventing the development of organic farming and squashing niches in which farmers' markets and local shops could survive or develop.[149] However, if the processed and packaged food especially associated with supermarkets is no longer seen as the source of the problems, the supermarkets themselves can go from part of the problem for food and climate change, to part of the solution.

This is certainly the view presented in the Strategy Unit report, which argues, predictably, that supermarkets exist solely because of consumer demand: 'expansion of multiple grocery chains is a long term phenomenon, driven by consumer demand for convenience, choice and a one-stop shopping experience.'[150] The report sees government improving consumer choices through supermarkets, not in questioning corporations' role in the environmental damage caused by the food system, and makes it explicit that processing food is no longer seem as the slightest bit problematic. In the future, we are told, 'consumers will be able to access healthy, low-impact food that fits their lifestyles, whether cooking from basic ingredients or buying a prepared meal.'[151] This is not restricted to governmental reports: Zero Carbon Britain, for example, see a role for food-processing and developments in food technology to satisfy consumers' desire for beef and dairy-like foods without the need for the cows.[152] Locating the problems caused by food production in the existential nature of meat and dairy products puts all food suppliers and distribution on the same level. Whereas in the food miles arguments, environmental concerns seemed a clear reason to oppose supermarkets and support small shops and farmers' markets, the greenhouse gas emissions caused by producing a piece of beef will be the same wherever it is sold. In this way again, the shift from processed foods to meat has an effect on the political nature of the arguments about food and climate change. It could be argued that it is not deliberate, but it is nevertheless profound.

The argument that the problem with food and climate change is people eating meat represents a fundamental shift in thinking about food issues, and places individual consumer decisions at the centre of thinking about climate change. It is this that enables the identities of problematic consumers to assume the importance that they have, and for fat people – and therefore by elision working-class people – to become responsible for destroying the planet. It wasn't the working classes' problem when it was the structure of the food system causing greenhouse gas emissions, but now it is their decision to eat hamburgers.

One obvious objection at this point is the connection between meat-eating and obesity. Being fat is clearly identified with being working-class, with all the accompanying stigma about lazy scroungers, drains on the public purse and so on that the identification entails in arguments about the fair distribution of resources. The line from obesity to meat-eating, on the other hand, is not an obvious one, since meat, particularly red meat, is not universally acknowledged as the major dietary *faux pas* of the overweight. Despite this, however, it is clear that the new focus on meat-eating as a problem for the climate is talking about fat people and working-class consumption.

The equation of meat and fat

While some authors do decry the consumption of food without nutritional benefits, like alcohol and sweets, when talking about the effect food consumption has on the environment, as we have seen, the main thrust of the case for the food system's contribution to climate change and world hunger is that certain foods, in particular meat and dairy products, have intrinsic properties which mean that they are unsustainable. This seems to be at variance with the idea of an obesegenic environment which encourages people, particularly working-class people, to eat junk food, since the production and consumption of processed food was one of the issues highlighted by the food-miles-type arguments, from which we are supposed to have moved away. While steak may be seen as diet food only in the context of

Atkins, would-be dieters are more likely to be advised to avoid fizzy drinks and crisps than they are simply to eschew meat and dairy. The fit between obesity and meat-eating is not a particularly natural one, underlined by Compassion in World Farming's admission in their recommendation for less meat-eating that 'it is impossible to be specific about the contribution of animal products to levels of obesity'.[153]

And yet, that there is a connection is an assumption underlying the bulk of contemporary discussion about the effect of meat production on the climate. Partly this is because arguments about the undesirability of meat from a climate point of view can pick up on views that eating a lot of it isn't optimal for your health either, and present them as a side-benefit which should encourage people to do the right thing by their planet and choose the vegetarian option. Thus it is that the arguments that individuals should reduce their meat consumption for the sake of the planet are often buttressed with claims about health, although these don't necessarily involve obesity. Pimental et al, for example, argue that cutting junk food intake and eating less meat would improve Americans' health as well as reducing fuel consumption,[154] while Compassion in World Farming base their recommendations for cutting meat-eating on dietary recommendations from the British Heart Foundation[155] and the Chair of the Department of Nutrition at the Harvard School of Public Health,[156] rather than thinking about obesity specifically.

These sorts of arguments create a connection between meat-eating and obesity without having to convince anyone that there is a causal relationship. In doing so by referring to respectable concerns about the health effects of eating a lot of meat they are the evidence-based tip of an iceberg of assumptions that meat-consumption, overconsumption and obesity are all effectively the same thing. It is noteworthy how, when diets are discussed from an environmental standpoint, meat-eating is presented as the same as over-eating. Meat-eating, it seems, causes obesity not because it is particularly fattening but because it is by its very nature unrestrained behaviour, and we all know that fat results from a lack of self-restraint.

The Meatless Mondays campaign provides a good illustration of how meat-eating and obesity are merged in thinking about overconsumption. As discussed above, the campaign was launched in June 2009 as a way of tackling the greenhouse gas emissions associated with meat,[157] but it clearly identified meat eating *per se* with overconsumption, and a vegetarian day with more restrained behaviour. Speaking on Channel 4 News on 15th June 2009 about the campaign, Paul McCartney explained the choice of Monday as the day for people to eschew meat not as one of alliteration but explicitly in terms of overconsumption: 'The idea is for people to give up meat for one day a week – Monday – when they've overdone it over the weekend, possibly'. The campaign's website continued with the presentation of the campaign as a way of devoting a day to 'healthy' behaviour, placing this even before climate change considerations: 'Our goal is to help reduce meat consumption by 15 percent in order to improve personal health and the health of our planet'.[158]

This mission statement did not make an explicit connection between meat-eating and obesity, but the 'About' page went on to list obesity among the various 'chronic preventable conditions' the risk of which could be reduced by going meatless for one day a week. The effect of the rest of the website was to make obesity a central, if not the central, issue for the campaign, with news feeds about dieting and obesity issues and promotion of the delights of different vegetables in terms which recall dieting advice. One recommendation of radishes, for example, highlighted not their taste, or use in different dishes, but that '[their] carbohydrate content fills you up fast while keeping you slim'.[159] The relevance of much of the site content to the central issue becomes clear only if obesity and meat-eating are seen as synonymous. What is under discussion is not really the choice of some food groups over others but the practice and recommendation of virtuous restraint: eating less than or differently from what you might wish because that is healthy and therefore morally desirable. In consumption terms, this is not entirely dissimilar to the restraint which Marx pointed out that workers were expected to practice in order to reproduce their labour at minimal cost to the capitalist, and which, as we'll see in chapter 4, was also identified by Malthus as the only thing standing between the poor and starvation.

The obvious rejoinder here is that creating a connection between obesity and unsustainable foods, even if it doesn't entirely stand up, is a strategy to improve public engagement. It is often argued, after all, that people have difficulty connecting climate change issues with the realities of their day-to-day lives, and that an issue which is simultaneously serious and distant is a public turn-off. Thus, the conflation of climate change and obesity in campaigns like Meatless Mondays is likely to be at least partly about providing additional motivation, in the same way that arguments about motivating cuts in consumer consumption discuss the advantages to the individual consumer, in terms of shorter working hours, improved quality of life etc. The argument that adopting a meat-free diet would be beneficial, not only to the planet but also to individual health, is clearly supposed to be a key element in persuading people to change their diets. In addition to the familiar idea that people are self-interested and need answers to the 'how will this benefit me?' question before they will do anything, it may also be assumed that modern Western consumers are well accustomed to messages telling them to abandon foods they like for the sake of their health, so dietary change for the sake of the climate message fits into a discourse which is already familiar.

It therefore becomes a kind of nudge theory, where small benefits to the individual are used as a way of persuading them to adopt behaviours that the government wants to see. It is possible that it may be effective with some people, although whether casting meat-eating in particular and Western diets generally as 'naughty but nice' is really a sensible way to approach issues of food and climate change is rather more dubious. The wider point however is that the equation of meat-eating with obesity closes the final link in the chain of reasoning that says that, while we used to think that the structure of the food system, the globalisation of food markets, the long supply chains for processed food and so on were responsible for food's malign effects on the climate, now we know that the real criminals are working-class people in the West who are choosing to eat too much and eat the wrong things.

If being responsible for climate change isn't enough, it doesn't stop there. The charge that working-class people's food choices are causing undue quantities of greenhouse gas emissions is in a sense a side issue to the greater charge: that they are simply taking much more than their share of the world's resources and that, if the world's population is to be fed, they will have to be restrained. That Western overconsumption is the cause of problems for the rest of the world is not a new idea, but the way in which it too became all about the working class is a relatively recent innovation, which needs to be explored.

3

Consumption, carrying capacity and class

The question of greenhouse gas emissions is not the only issue for food production. There is a growing sense that the demands of feeding the world's population are too much for the world's resources: the efforts to do so are not only causing environmental destruction, but will ultimately fail. It's often said that climate change is a difficult issue for people to engage with, especially in the developed world, as it doesn't appear to have concrete, immediate applicability to their everyday lives. Warnings of coming food shortages have a long history, but in the last few years this argument appears to have gone from an abstraction to an urgent reality, set as it is against the background of surging prices in basic goods across the world.

In 2007 and 2008, food prices spiralled worldwide The cost of Thai rice went up by more than 200% between 2007 and April 2008, with increases of 106% for oils and fats, 48% for dairy products and 88% for cereals in only twelve months between March 2007 and March 2008.[160] Overall, the International Monetary Fund (IMF)'s index of internationally-traded food commodities rose 56% between January 2007 and June 2008.[161] The high prices of course hit the poorest people hardest, and as people were forced to cut their consumption of meat, dairy, fruit and vegetables just to get enough grain or rice to eat, malnutrition soared: it's estimated that the number of malnourished people worldwide increased by 75 million in 2007, and another 40 million were added to that total in 2008.[162]

With 848 million people suffering malnourishment in 2006,[163] these increases by themselves might not necessarily have impinged much on Western public consciousness. That the food price rises of 2007-08 were headline news in the developed world must be in part because the effects were felt here as well, if not so catastrophically. But it is also testament to the fact that, if people were starving, they were not starving quietly. There were demonstrations and strikes around the world as people protested at the cost of staple goods soaring beyond their reach. In January 2007, 70,000 people demonstrated in Mexico City at the doubling of tortilla prices, and protests followed during 2007 and 2008 in, among others, Morocco, Mauritania, Senegal, Indonesia, Burkina Faso, Cameroon, Yemen, Bangladesh, India, El Salvador, Haiti, South Africa and Italy.[164] In Egypt, a strike by textile workers at Mahalla against rising food prices became a mass demonstration when the strikers were attacked by security forces,[165] and the then-President Mubarak was forced to mobilise the army to produce and distribute subsidised bread in an attempt to contain the protests, which were a preparing ground for the 2011 revolution.[166] Food price rises returned in 2010, as between June and August, wheat prices increased by 50%,[167] described by HSBC analysts as 'the most dramatic rise for more than 30 years',[168] and there were also significant rises in the prices of pork and cocoa.[169]

Why food prices soar

The reasons for these price spikes range from shortages caused by harvest failures, to shifts from food to biofuel production and market speculation in commodities. Which of these factors is the most important depends on who you ask, with seemingly every report on the 2007-08 crisis, and the second wave of price rises in 2010, giving a different emphasis. Much of this is unsurprising, as those at risk of being seen as responsible for causing widespread hardship seek to shift the blame. So, for example, Gruma, Mexico's chief tortilla producer, defended itself against charges of hoarding supplies by blaming the rises in corn prices on increasing US demand for ethanol, made from corn,[170] while the UK government's Strategy Unit defended the financial markets in 2008 by explicitly denying any possibility that speculation could be behind the price rises.[171] It was noted in August 2010 that even the World Bank had performed an 'about-turn' on the principal cause of the 2007-08 crisis.[172] A World Bank policy research paper written in July 2008 blamed 'biofuels and the related consequences of low grain stocks, large land use shifts, speculative activity and export bans' for 70-75% of the food price rise, with the other 25-30% the result of high energy prices.[173] In this analysis, speculation was the result, not the driver, of high food prices. However, two years later, after the 2008 crash, another World Bank paper concluded that the contribution of biofuels was overstated, but that financial speculation was a significant cause of the 2007-08 price rises.[174]

There is clear evidence that speculation in commodities was an important cause of the food price rises in 2007-08. As the World Development Movement (WDM) explain in their report on the food crisis, *The Great Hunger Lottery*, market speculation in food and other commodities has been happening for a century, through trading futures contracts. Futures started as a way for farmers to protect themselves against prices falling in the time it would take for their produce to grow and be brought to market, by setting the future price for which the produce would be sold when it was ready. In the nineteenth century, the prices set in these futures contracts were related to what traders thought the level of demand for the produce would be, but in the early twentieth century, futures contracts began to be traded on stock markets as if they were just another financial instrument, so the prices often bore little relation to agricultural realities.

The early years of this century saw a huge increase in trading in food derivatives, as a result of relaxation of financial market regulation. Between 2002 and 2008, the number of derivative contracts in commodities increased by 500%, and speculators held 65% of the contracts in maize, 68% of soybeans and 80% of wheat.[175] This meant that the prices in these and other commodities were extremely volatile, so, for example, spring wheat prices leapt by 25% on the US markets on a single day in early 2008.[176] As WDM conclude, it is difficult to explain these sorts of swings except as the result of speculation,[177] which as well as causing price swings, also clearly pushed overall prices higher than they otherwise would have been. Even before the worst of the food crisis, in 2006, Merrill Lynch admitted that in their estimation, speculation in commodities was making the prices at which they were trading 50% higher than if they were based only on supply and demand.[178]

The idea that there is a fundamental problem with global supplies has been used as a counterargument to the evidence of the effects of speculation on food prices. The UK government Strategy Unit, for example, stated in their 2008 report on food that the financial speculation argument was 'difficult to substantiate' (even if this was so in 2008, the WDM and World Bank reports show that it is no longer), and added that 'the basic mismatch between demand and supply, and the resulting tightening of markets, is the fundamental issue'.[179] While a continuing commodity price 'boom' would of course be extremely bad news for most people in the world, it would be very profitable for investors who could benefit from continually rising prices, and for the commodity dealers. Goldman Sachs, for example, argues that increases in food prices are caused by structural issues and will continue whatever the market does. The fact that in 2009 it made around $5 billion from commodity trading[180] should suggest that this view is not entirely disinterested. However, while the food crisis of 2007-08 gave new impetus to fears that global production was up against its natural limits, it was not the origin of this argument. We have, it seems, been reaching the natural limits of food production for some time.

An important aspect of understanding food's contribution to climate change is that climate change will have a serious effect on food production. As a result of more extreme weather events it will become more difficult to grow crops or raise livestock in many parts of the world. Drought in sub-Saharan Africa, floods in Bangladesh and Pakistan are only indications of what the effects of global warming may be, and these effects will not be localised. The global market in food means that crop failures in one part of the world could affect food supplies throughout. Crop failures and falls in dairy production because of droughts in Australia and New Zealand are often cited as a factor behind the 2007-08 crisis, while the record heatwave and drought in Russia and the Pakistan floods provided a background to food price rises in 2010. While the arguments that the food price increases in 2007-08 were driven largely by speculation is, I think, convincing, the climate conditions limiting food production in various parts of the world in 2007-10 could be taken as a sign of things to come, and as evidence of how food production not only affects climate change, but is affected by it.

The problem with biofuels

The need to respond to climate change can also create constraints on the use of land for food production. Deforestation is a significant cause of global warming, so forest clearance to create more arable or grazing land is clearly not a climate-friendly option. There is also the question of competition for land use from biofuels. As noted above, diversion of arable land to grow crops for biofuel production has been singled out as a cause of the 2007-08 food price crisis. Although it seems unlikely that a supply constraint caused by biofuel production could have caused the wild price swings seen in the crisis, it seems evident that increasing production of biofuels could have an effect on the amount of crops available for food. Biofuels were the source of the highest increases in demand for crops in the years leading up to the food price crisis: in December 2007, the International Food

Policy Research Institute reported that demand for cereals for biofuels increased by 25% between 2000 and 2007, compared to a 4% increase in demand for cereals for food and a 7% increase for animal feed, and that in the same period, the use of corn for ethanol production in the US went up by 250%.[181]

Replacing fossil fuel use with biofuels may seem an impeccably green idea, but in particular because of their impact on food availability, biofuels have been seen as increasingly problematic. 2007 has been called 'the year the world woke up to the significant climate, social and human injustice impacts of this new energy technology being foisted on the world', with calls for moratoria on biofuel expansion.[182] This growing awareness that production of biofuels for use in the developed world could be taking the place of food crops for the poor of the developing world has been countered with promises of 'second generation' biofuels with 'new and improved technologies'.[183] Whether these would be able to create genuinely sustainable biofuels on a large scale is however unlikely,[184] and the source of the material for particular biofuels is often obfuscated. The story of the UK government's biofuel requirement is instructive on this point. In April 2008, it introduced a requirement that 2.5% of all road transport fuel had to be biofuels, as part of an EU directive requiring 10% biofuel use by 2020, only to discover that a measure which it had hoped would bolster its green credentials actually brought it considerable criticism, not helped by the timing of the requirement just as concern at food prices was at its height. The government's response was to support second generation biofuels, but to its embarrassment, four months later it had to admit that it could not trace the source of 80% of UK biofuel imports and therefore could not rule out that they were being produced from potential food crops.[185]

The biofuels issue is of course more complex than simply the question of the distribution of crop land between growing fuel and growing food. At base it is an issue of social justice, as the malign consequences of increasing biofuel production – deforestation, food insecurity and so on – proceed not simply from changes in land use but from the replacement of small farmers with multinational biofuel companies. The problem is not just biofuel production, but the structure of global agrofuel production, in which farmers in the developing world are either forced off their land or driven to switch from producing food for themselves and local people to their own cash crop production.[186] However, it is clear from recent reports like that of Zero Carbon Britain that even when we are talking about planned large-scale biofuel production in the developed world, biofuels and food production are essentially counterposed. Even removing the concerns which make biofuel production in the developing world such a social justice issue, there remains the assumption that our food production is too resource-intensive to allow a biofuel response to the problem of fossil fuels. Zero Carbon Britain, indeed, argue for a drastic cut in the amount of land used for food production, 'to release land for other decarbonisation purposes'.[187]

The impression given by the land use section in the Zero Carbon Britain report is not of a fundamental shortage of land for food production use: the calculations here provide food, if not resource-intensive food, for the UK population using only 29% of current agricultural land.[188] This might imply that even a small island like Britain has plenty of agricultural land to spare. However, at the base of the arguments about the food crisis is the idea that there is a fundamental resource constraint when it comes to food production, exacerbated by climate change and potential measures to respond to it, but existing separately from the effects of global warming.

Fundamental resource constraints

The fundamental resource constraint is ever-present in discussions of food and climate change. The problem facing livestock farming, according to the FAO, is nothing less than 'the planet's finite natural resources'.[189] For Garnett, it is 'that there is only so much to go round',[190] while the question of the finite planet is made most explicit in a 2008 Chatham House report on the future of food, which centres around a discussion of whether 'the widely feared fundamental limit to global food production' is reached or delayed.[191] Awareness of 'the insight that nature limits us'[192] or the fact of 'the finite resource base and the fragile ecology on which we depend for survival'[193] permeates much of the recent literature on food production. On one level, of course, the fact that the planet is not literally infinite and none of its material resources will go on forever is a truism; the point is that the fundamental limits which logically must exist are becoming relevant, because in this view we are nearly at them.

The effects of climate change, as set out above, clearly play a role in limiting possible food production, as do attempts to respond to climate change (whether helpful or otherwise), such as avoiding deforestation or producing biofuels. These are not, however, the only stresses viewed as pushing against planetary limits. The FAO summed these up as 'the additional demands [placed] on the environment by a growing and wealthier population',[194] a conclusion reflected with varying stresses by other writers. For some, the fact of increasing population is the main problem, although not always in such apocalyptic terms as this particularly noxious recent example: 'A global hurricane of consumption from these rising populations [in Bangladesh, China, India, Indonesia, Nigeria and Pakistan] is gathering force as it sweeps through each generation'.[195] On the other side of the formulation, the rising consumption of the middle classes in India and China has become a well-known fact for many, such that it is frequently repeated in media discussions of food and resource constraints. In April 2008, for example, a *New York Times* editorial cited 'the march of the meat-eating Chinese – that is, the growing number of people in emerging economies who are, for the first time, rich enough to start eating like Westerners' as a cause of food price rises',[196] while a month later, *The Guardian* was reporting from China 'where rising demand for meat from a growing middle class is destabilising world food prices.'[197] In 2011, a *Nature* article by 21 authors summed up this thinking when it argued that food production would need to double to keep pace with population increases and projected dietary shifts towards more meat and dairy consumption.[198]

Within the current system, there are two obvious conclusions which could be drawn from these views of the problems of food, the less unacceptable of which is that, if there is more strain on food resources, those who are currently using the lion's share should reduce their consumption. The other, particularly with regard to population growth, will be discussed below, as its relationship to both food and class is not as straightforward as it might first appear. For those who are not prepared to argue that Indian and Chinese middle-class people should be prevented from increasing their meat consumption, the only option seems to be to call for less resource-intensive eating patterns in the West. If people in India and China are eating 'like Westerners', Westerners themselves will have to eat in a less 'Western' fashion.

As with the arguments about food and greenhouse gas emissions, the argument about resources also places red meat and dairy products at the top of the hit list, as the most resource-intensive food products. The FAO in their 2006 report highlighted the contributions of the livestock sector to problems of land degradation, deforestation, water use and pollution of water supplies, the latter particularly important given the effects of climate change on rainfall patterns and water supplies in many parts of the world.[199] However, in the context of a crisis in human food production, the most important issue about meat appears to be the sheer amount of crops grown not to feed humans directly, but to feed cattle. In his introduction to Compassion in World

Farming's 2004 report, Jonathan Porritt estimated that by 2050 cattle would be consuming the same amount of crops as four and a half billion people – equivalent, he noted, to the entire world population in 1970.[200] Much of the water use associated with livestock is also not drunk by the cattle directly, but used to irrigate the crops grown to feed them.[201] This is considered an inefficient use of these scarce resources because not all the calories eaten by the cows are passed on to the humans eating their meat or drinking their milk. In approximate terms, one kilogram of chicken meat would take two kilograms of feed to produce, while one kilogram of pork would need four kilograms of feed, and beef, seven kilograms.[202] If humans ate 'lower down the food chain',[203] seven kilograms of crops could be turned into seven kilograms of human food.

Changing meat versus changing the system

On one level, this, like the arguments about food's contributions to greenhouse gas emissions, can appear as a technical problem, with solutions on a similarly technical, as opposed to systemic, level. This is not to say that there would not still be considerable debate about the precise nature of the problem and the way to solve it. The FAO's conclusions for example – that livestock farming needs to become more intensive, essentially moving away from small, subsistence farming to farming by large conglomerates – are supported by some writers such as Tara Garnett for the Food Climate Action Network, but have also been hotly disputed.

Simon Fairlie, for example, argues that the answer is re-ruralisation of what should be a much-reduced livestock sector, rather than increasing the trends towards that industrial, intensive production which is the source of the problems of livestock farming.[204] However, it is clear that even though the issues are frequently expressed in the technical language of so much grain consumed by cows rather than humans, or so much methane emitted, they should be understood as existing at a much deeper, more fundamental level; as being, in fact, problems of the system and not simply of details of our diets. The comment that we should eat 'lower down the food chain', quoted here from Compassion in World Farming but an often-repeated concept, is instructive.[205] It seems to express far more than that people in the West should eat different animals, or that farmers should feed them differently; rather, it calls for a significant shift in the way we view food consumption in the developed world, and indeed the food consumption of the middle class in the developing world.

These two levels to the problems and suggested solutions for food – one an essentially technical problem, responsive to changes implemented within the current system, and the other at the systemic level – reflect the debates around many issues in climate change. For many green campaigners, the worst insult you can level at any suggested action is to call it a 'techno-fix'; that is, a putative solution to a climate change problem which tries to achieve its aims without changing how the system which caused the problem works in any meaningful way. The most obvious techno-fix proposals of recent years are things like the space mirrors or reflective dust which the US recommended to the IPCC in 2007,[206] but the concept is not restricted to the outlandish. The techno-fix is any measure which changes the technology rather than the system which uses it, whether that technological change is the replacement of petrol with biofuels or coal-fired power stations with wind turbines. The argument is that any such technological changes are an attempt to 'decouple' production from the ecological damage it causes, and there is considerable debate about whether that would be possible, or whether positing that it might be possible to have production without its material consequences is simply an attempt by interested parties to maintain the *status quo*.[207]

In the context of the arguments around food, advising people to eat less meat, or at least to eat chicken instead of beef, may not seem a very technical techno-fix, but it can be seen as a solution on that level. The suggestion for example that food scientists could develop 'fake' beef and other meats, so that people did not have to lose their taste for them, does demonstrate a line of thinking which looks to scientific solutions rather than system changes.[208] The alternative is to see damage to the ecosystem caused by food as part of the system of food production, so that change also has to happen at that systemic level: changing the system, rather than tinkering with it, as the techno-fix tries to do. This latter is an approach taken by much of the literature discussed here about food and climate change, and fits the issues of food into a much wider, long-standing debate about overconsumption and its effects on the planet.

In this respect, the natural limits which appear in debates about food and climate change are not simply the limit to the amount of food which can be produced on the planet, but an expression of the fragility of the biosphere as a whole. Climate change here is not just a factor in setting limits to food production, by for example making it inadvisable to clear forests to create more agricultural land, but is evidence that the natural limits are already being breached. The argument is that our consumption levels are such that we are failing to live within our environmental means. As far as food is concerned, the evidence that consumption of 'inefficient' foods in the First World has a direct linear relationship with hunger and poverty in the Third World is not actually particularly good. A computer model developed by Compassion in World Farming found that a 50% shift by Western consumers to less resource-intensive food would reduce childhood malnutrition in the developing world by a measly 2.8% by 2020.[209] However, the assumption remains that global consumption is a zero sum game, in which a fair share would be distinctly less than the abundance associated with the West: 'the overconsumption of the billion or so who consume far more than their basic needs and, it is reasonable to assume, contribute directly or indirectly to the underconsumption of the impoverished billion'.[210]

It is this definition of the problem with food as one of Western consumer behaviour which makes the connection between food and overconsumption more generally. The idea that modern Western societies in general, and individuals within them, consume too much is frequently a given in discussions of environmental problems. Anyone reading broadsheet media in the UK probably encounters it several times a week. Because it has become an accepted truth in many circles, its intellectual origins are often not explained, but this doesn't mean that it doesn't have them. The idea that consumption is problematic has a considerable pedigree.

Overconsumption and the steady state

The argument that overconsumption is particularly problematic takes its inspiration among others from John Kenneth Galbraith, in his 1958 work *The Affluent Society*.[211] Galbraith argued that the unfettered growth which he saw the US government pursuing in the 1950s was not beneficial for society and, despite statements to the contrary, did not alleviate poverty. In his 1976 introduction to the third edition, he explained how the question of poverty in areas like the mountains of West Virginia, eastern Kentucky and Tennessee was his inspiration for the work, for which his working title was 'Why People Are Poor',[212] and how it was intended as a refutation of the 'trickle down' theory beloved of post-Second World War free market economic thinkers. Although he had been a Keynesian in the 1930s and 1940s, after the Second World War Galbraith saw other Keynesians adopting what he saw as an obsession with production for the sake of production, and a theory of constant economic growth which had the effect of starving public services while building up private wealth for the few. If, he argued, production was no longer treated as an unequivocal good to which all other interests could be sacrificed, society could develop 'a more relaxed and compassionate view of output without putting individuals to hardship'[213]; a world with shorter working hours, lower unemployment and better public services.

Since Galbraith wrote this, such ideas about the undesirability of continuous economic growth have become important for thinking about environmental problems, underlying for example the steady state theory. This theory, developed by Herman Daly, views economic growth as the structural problem that causes environmental damage. Using John Stewart Mill's theory that the economy would expand to a certain point and then stop in a 'stationary state',[214] it theorises an economy with a stable size, neither growing nor going into recession, with constant stocks of labour and of capital, so that apparently the flows of goods and services produced would also be constant. Economic development, in the sense of change in what was produced or the way in which it was done, could still occur, but it would be change rather than growth, since in this theory, the constant labour and capital stocks would prevent it.[215] Because such a society would have limited production and population, the theory holds that its effect on the environment would also be limited. While manmade capital used to be the limiter for growth, henceforth it would be natural capital.[216] In fact, society would have to live within its natural means to avoid further environmental harm, tracked by, for example, the concept of 'ecological debt' developed by Andrew Simms.[217]

Economic growth in this discussion is so malign because it can only be achieved through increases in production and consumption, with all the depletion of natural resources and environmental damage that entails. The steady state theorists are not necessarily anti-capitalists: Herman Daly was a World Bank advisor, and he and others were at pains to stress that the trappings of capitalism could continue, imagining for example a steady state with a stock market.[218] The problem of how the capitalist system could work without growth is therefore a real problem. One suggestion is that material production could be replaced with capitalist economies based on providing services, which in this idea would not have the same material effects on the environment, but not everyone agrees that this decoupling would be possible. Daly himself was sceptical about the idea that labour could replace material inputs in production, pointing out that if you're building a house and you run out of wood, you can't make up for it by hiring more carpenters. When Tim Jackson reviewed the material/immaterial production problem in his 2009 *Prosperity without Growth?*, he concluded that 'the capitalist model provides no easy route to the steady state'.[219]

Whether or not it is believed that immaterial services could replace the economic effect of material production, it remains the case that for these theorists, material production and consumption is the source of our environmental woes, just as for Galbraith it was the cause of social and economic problems. Production and consumption are obviously connected as two parts of the same process: production without consumption would be distinctly unprofitable, whereas consumption without production would quickly run out of things to consume. However, identifying the process as the cause of ecological destruction necessitates also identifying which out of production and consumption is the key driver of it.

For the writers on the steady state, it is consumption which is clearly the most important side of the equation, with recent works treating it rather than production as key to issues of economic growth and environmental problems.[220] This is in line with general received wisdom that consumption drives production: consumers conceive a need for products, and so the market provides them. As any marketing textbook will tell you, the way to a profitable business is to 'determine the needs, wants and values' of your target market, and develop products to satisfy them. Producing the products you want, then cajoling unwilling consumers to buy them, is not a highly-regarded strategy. Despite the all-pervading nature of this assumption, it isn't necessarily correct. Marx for example argued strongly that despite bourgeois economists' views, production leads consumption, not the other way around. Workers' ability to consume is important to the capitalists producing the items for their consumption, but the production is not created by their demand, which in fact would never be sufficient to generate profits from production.[221]

Galbraith was not a Marxist, and his attitude towards Marxism is difficult to interpret, summed up as it is in the immensely quotable but bizarre comment that 'the man who argues with a Marxist has always been assaulting a rock fortress with a rubber flail'.[222] He does however seem to have agreed with Marx on the relation of production to consumption, stating that 'production only fills a void that it itself has created'[223]; in other words, that advertising created consumer wants which would not otherwise have existed in order to sell what capitalists chose to produce. His argument was that production, and ever-increasing production, was not naturally pre-ordained, as he thought some contemporary commentators perceived it. He saw the excuse that 'it would damage production' being used to deny more public spending on health, education, public infrastructure and other elements of social welfare, and the main goal of *The Affluent Society* was to argue that this pre-occupation with production was both harmful and unnecessary.

As a result, the solutions put forward by Galbraith were largely policy ones, like shorter working hours, more public works and higher taxes. In contrast, arguments since around the problem of overconsumption have tended to focus on consumption-related solutions, on the assumption that if consumer demand changes, production with all its environmental consequences will automatically follow. This has meant that they have had to examine consumption in some detail. The steady state as an economic theory can to some extent treat all consumption as one category, but if we are setting out to reduce consumer demand for products, this demand has to be subdivided. It is clear that for these arguments some consumption is much more problematic than others. Further, not all consumers are the same. An argument which on one level seems to approach consumption from all sections of society as potentially the same seems on closer inspection to be very concerned about the class of consumers under consideration.

Consuming and class

Galbraith himself is vague on the question of his definition of consumption, nor does he clarify which class he is imagining doing the consuming. It is fairly clear that consumption for Galbraith is consumer consumption as might be understood in the general, popular sense: buying the products of new technologies. The modern object of this consumption would be iPhones, LCD TVs and laptops; in a 1950s context, we should probably imagine more items of household equipment like fridges and vacuum cleaners, and, of course, cars. Galbraith's discussion of the irrationality of consumer wants, spurred on by advertising, conjures a familiar image of 'must have' products which fits very well with a modern understanding of consumer consumption as indulgence in products which, while pleasant and desirable, are unnecessary for life.[224] But this also raises the question of the identity of the consumers.

The question of class and consumption is a complex one. While at its basic level, consumption of 'consumer' goods like high-end electronic gadgets implies a certain level of material wealth, there are also common ideas which single out working-class consumption and which imply that consumption of consumer goods is as much a working-class as it is a middle-class behaviour. For example, as we've seen, a familiar and egregious argument criticises working-class people for consumer consumption, either for wasting their dole money on widescreen TVs and Sky subscriptions, or complaining that if they can find the resources to acquire TVs and Playstations, they can't really be poor. It's an idea at the heart of the neo-liberal view that there is no such thing as poverty in the developed world, as set out for example by Margaret Thatcher in 1978: 'Nowadays there really is no primary poverty left in this country. In Western countries we are left with the problems which aren't poverty. All right, there may be poverty because people don't know how to budget, don't know how to spend their earnings, but now you are left with the really hard fundamental character-personality defect.'[225]

This has particular application to children, with the concept being that middle-class parents know how to educate their kids while working-class parents simply pacify them with computer games and expensive trainers, and carries with it the understanding that anything consumed by working-class people becomes devalued in middle-class eyes. However, these are essentially criticisms of working-class people attaining access to goods that might once have been restricted only to the middle classes. They do not contradict the fundamental point that consumers are those who have the resources to be able to consume, and the more expensive the goods under discussion as problematic consumption, the wealthier the consumers are likely to be.

Galbraith's picture of private wealth and public poverty seems in places to imply truly serious wealth, so the impression is that the articles of problematic consumption would be luxury limos and heated swimming pools, with possibly a new vacuum cleaner for the maids. However, he may have had consumption by much poorer consumers in mind. In the introduction to the original edition of *The Affluent Society*, Galbraith said that society was living with economic and social ideas developed in a world where poverty was the norm, and the challenge was to adapt them to a new world where most people were affluent.[226] In light of this statement, the problematic consumption under discussion in the book would not be the consumption of the wealthy elite, who had existed in the previous world of general poverty, but of the newly un-impoverished majority. The US in the 1950s saw a great increase in the accessibility of consumer goods to working people. It seems that the consumption Galbraith was talking about was not so much that of the wealthy *per se*, but the threat that behaviours once limited to the wealthy were becoming available to the wider population. The similarities between this and the current arguments about middle-class Indians and Chinese adopting Western diets are not coincidental.

The key idea here is that of unnecessary consumption: consumption which the previously-impoverished workers engaging in it might not recognise as such, but which is damaging to their long term interests if it holds back the development of public services and keeps them working long hours to earn the wages to afford it. Galbraith effectively divided the consumption of ordinary Americans into two categories: that of their basic needs, and luxuries, many of which in his opinion they did not really need. This was not a distinction invented by Galbraith himself, but while he probably was not influenced directly by Marx, the division he makes is reflective of Marx's discussion of workers' consumption.

Definitions of consumption

Marx points out that the absolute lower limit for wages is the cost of reproduction of labour. If workers don't earn enough to support themselves and their families, this not only leads to misery for them but impedes the development of the next generation of workers, so it is ultimately necessary for the capitalist to ensure that their workers' children are minimally, if not particularly well, fed. However, Marx also points out that in practice it is not always possible for employers to keep wages at the level of simple reproduction, so workers were sometimes able not only to increase their consumption of necessities, but also 'enjoy momentarily articles of luxury ordinarily beyond [their] reach, and those articles which at other times constitute for the great part consumer "necessities" only for the capitalist class'.[227]

This point was later developed by Rosa Luxemburg, when she described how workers' struggles can achieve not only the momentary increases in living standards described by Marx, but also advances in the customary standard of living above the physical minimum for survival. In this way, successive struggles can establish a cultural and social standard of living, below which wages cannot fall without provoking severe resistance, and which can become accepted by employers as a *de facto* minimum wage level.[228] An example of this process could be seen in the way in which consumer goods like fridges and televisions have become progressively more accessible to working people over the last fifty years, going from luxuries, to necessities, to items which are assumed to be part of a basic standard of living. The disparaging comments sometimes levelled at working-class possession of such items as showing that they are 'not really poor' can be seen as railing against the process identified by Luxemburg. Certainly, attempts to reduce wages to a level at which a fridge would be an unaffordable luxury would be met with considerable resistance in most industries.

As the result of successful workers' struggles, the advance of the customary standard of living can be seen as a defeat for the capitalist class, in that it makes it more difficult for them to reduce wages below a certain level. However, while it might mean that employers incur a higher wages bill, it would also be a boon for the fridge manufacturers, since everyone else's workers would provide a new market for their fridges. Marx points out that to the capitalist, the total mass of workers excluding his own appear as consumers: since capitalists always regard wages as a loss, rather than a gain, the unachievable ideal for any given employer would be for everyone else's workers to be generously paid, but for the wages of his own to be held at the basic level of reproduction.[229] The credit bubble which burst in 2008 is illustrative of an attempt to get closer to this capitalist nirvana, as corporations shifted jobs to parts of the world where wages were low, while credit substituted for earnings in enabling workers in the First World to continue to provide a market for consumer goods.

It will be apparent from this discussion that for Marx, articles of production were divided into two categories, according to how they are consumed, or not, by the workers.[230] The first category he called 'articles of consumption', comprising items required to fulfil the workers' basic needs, while in the second category were the 'articles of luxury' to which the workers could only sometimes win access. In the light of the debates about overconsumption, it's important to note that for Marx, the definition of basic needs was not set by experts, bourgeois opinion or research into human physical requirements, but by workers themselves. His definition of workers' basic requirements was those items without which workers themselves would feel unable to live. His comment on items of which some modern commentators might question the necessity was that they were included in the category of articles of consumption 'regardless of whether such a product as tobacco is really a consumer necessity from the physiological point of view. It suffices that it is habitually such.'[231]

As we have seen,[232] Marx was scathing about the ideas circulating in the nineteenth century of the desirability of restraint and self-denial for the workers; the argument that they should limit their consumption as far as they could and hence build up savings: 'Society today makes the paradoxical demand that he for whom the object of exchange is subsistence should deny himself, not he for whom it is wealth'.[233] He pointed out that this demand was effectively that the workers should limit anything in their lives which did not consist of working, and hence generating more surplus value for the capitalist. They should 'maintain themselves as pure labouring machines' and even build up enough savings to pay for their own wear and tear.[234] In contrast to this, 'the worker's participation in the higher, even cultural satisfactions, the agitation for his own interests, newspaper subscriptions, attending lectures, educating his children, developing his taste etc, [is] his only share of civilisation which distinguishes him from the slave'.[235] From the point of view of modern ideas of the unnecessary and frivolous nature of much consumption, this might appear a list of rather highbrow pursuits, but that was clearly not how it was meant. These were activities in which workers might participate if they were able, and which, we might note, may have seemed devalued in middle-class culture had sufficient numbers of workers managed to do so.

The difference between Marx and Galbraith on this is that for Marx, workers' consumption, when possible, of the articles of luxury usually denied them, was also an unalloyed good thing, while for Galbraith, it was not only the source of social problems but the result of consumer desires which they did not really feel. Although for Galbraith, production was not demand-led, he clearly did view the ability of many working people to access things from Marx's second category of 'articles of luxury' as a problem. Rather than improving their living standards in the way that Marx described, Galbraith felt that some needs for consumer goods were unreasonable and undiscerning: the consumers did not distinguish between important needs and minor whims and did not know what they really wanted.[236]

The idea of unnecessary, ill-considered consumption has clearly been an extremely influential one, even as the definition of consumption has developed. Some of the more recent discussions of overconsumption have adopted Galbraith's understanding of consumption as consumer consumption; so, for example, the World Wildlife Fund's 2008 report on the environmental movement discusses motivating pro-environmental consumption in terms of consumer decisions specifically to buy a hybrid car rather than an SUV, and with a general focus on electricity use and electronic goods.[237] Whether consciously or not, this follows Galbraith's definition of the nature of problematic consumption as an issue of articles of luxury, as Marx defined them, and not articles of basic consumption. However, in much of the literature on overconsumption, the definition of excessive consumption has clearly changed.

The changing face of overconsumption

One of the earlier recent writers on overconsumption to look at the different possible definitions of consumption was Paul Stern. In a paper published in 1997, he outlined the popular definition of consumer consumption as 'the purchase decisions of households and what they do with their purchases', associated with problems like plastic waste, traffic, consumerism, advertising and shopping malls.[238] This is clearly the consumption of articles of luxury as criticised by Galbraith, although the definition outlined by Stern changes the focus from the act of producing the articles for consumer consumption, as in Galbraith, to their packaging and distribution. It is difficult to identify why Stern put the definition in precisely this way, but it is possible that he was reflecting a general sense that, as in the food miles arguments, the problem was not with the items themselves, but the system of industrialised production and globalised markets. In any event, the popular understanding of consumption seemed insufficient to Stern, who argued for a much more far-reaching definition, which would, he argued, enable people to appreciate human effects on the biosphere: 'human and human-induced transformation of materials and energy'.[239]

This definition clearly moved consumption out of the realm of articles of luxury and opened the concept of overconsumption to any form of consumption at all. Stern's definition does not single out the basic needs category of consumption; his principal target in arguing that the consumer definition of consumption is too narrow appears to be the public sector and, by extension, other non-domestic consumption.[240] However, this is not true of subsequent definitions of consumption. Thomas Princen, for example, followed Stern in adopting an environmental definition of consumption, but felt that Stern equated consumption erroneously with materialism.[241] His 'ecological' definition of overconsumption – 'consumption which undermines a species' life support system'[242] – is actually quite similar to Stern's, but his comments on the consequences of this definition demonstrate the direction in which the overconsumption arguments were moving.

For Princen, consumption over basic needs is always harmful. Galbraith thought that consumers could not always tell when they really wanted something; Princen implies effectively that they should never want anything at all. He identifies a goal of policy-making to identify instances where individuals were harming themselves and the environment, because these should be the easiest behaviours to change: 'win-win, no regrets policies that simultaneously produce improved human welfare and reduce ecological risk to humans' life support system'.[243] These opportunities would seem to be legion, as Princen ends the article with the comment that the world could 're-embrace such notions as thrift, frugality and self reliance'.[244] It is clear that if Princen did not consider his definition of overconsumption to include basic needs, those basic needs were defined pretty narrowly, without much room for those items which, although not physiologically necessary, are habitually considered such.

Princen is one of the most overtly right-wing of the recent writers on overconsumption, as shown by his argument that social change to address climate change was possible because 'if former communists can embrace free markets', anything can happen.[245] His vision of a society of self-reliant individuals is a far cry from Galbraith's plea for better funding for public services. However, while the packaging of his arguments may be different, Princen was not an outlier but firmly part of the development of the definition of overconsumption. It is clear from the most recent literature on overconsumption that the increasingly-accepted definition is one which can include food. This is demonstrated, for example, by Alcott's 2008 explanation of the 'sufficiency strategy', a plan for lowering greenhouse gas emissions and living within humanity's resources, in food terms. Achieving reductions in emissions by increasing efficiency is, he says, like not overfilling the kettle when making coffee. However, sufficiency means not having the cup of coffee at all.[246]

The coffee in this analogy would certainly have been included within Marx's definition of basic needs. It may not be strictly physiologically necessary for life, but it can seem so, particularly first thing in the morning. The fact that Alcott could use food-related consumption as his parable for the entire sufficiency strategy is a striking indication of how the definition of overconsumption shifted from Galbraith's understanding of consumer consumption of luxury articles to include elements of basic needs. He also explicitly includes some foods in a list of 'common ostensive definitions of dispensable or at least negotiable consumption': large houses, air travel, SUVs, cosmetics and meat.[247]

Alcott does not explain the basis on which these particular things are regarded as overconsumption. The arguments against air travel and SUVs are clear on greenhouse gas emission grounds, and large houses could be included in this on the basis of increased energy required for heat and light, although this is more an issue of insulation, choice of heating system and turning lights off than of square footage itself. This would imply that meat is included in the list on the grounds of its high emissions and inefficiency in terms of cereal and water use, as discussed above. Its equation in this way with air travel and SUVs is demonstrative of its high place in the chart of climate damagers, ahead, for example, of ordinary cars and lorries. However, the inclusion of cosmetics in the list of overconsumption implies that greenhouse gas emissions are not the only factor being used here to find items of consumption worthy or unworthy.

The cosmetics industry, while not especially low in emissions, is also not associated with high greenhouse gas emissions in comparison with production of other marketed, packaged products. Its inclusion in a list which does not also include things like TVs and mobile phones makes little sense, if this is indeed a list of the worst-emitting forms of consumption. However, it could be argued that cosmetics are not necessary for life, and might seem to someone who (presumably) does not use them as easier to dispense with than other sources of emissions, the lack of which might seem to have a more significant effect on lifestyle. In this light, Alcott's list of items of dispensable consumption seems not to be a list of products whose particularly high emissions make them especially unsustainable, but a selection, from the wide category of items whose consumption causes greenhouse gas emissions, of those things which he thinks we don't need. Thus the definition of overconsumption here still effectively excludes basic needs, but basic needs as externally defined; not, as in Marx, defined by those doing the consuming.

Implicit in the concept of unnecessary consumption is the idea that some consumption can be actively harmful, not only to the planet, but to the consumers themselves. Writers on overconsumption have sought to develop the idea that cutting consumption would be in the direct interest of consumers, as less consumption would make us happier. This is sometimes presented as if an argument for 'downshifting': we work long hours so as to be able to afford consumer goods, and could choose to work less and have less.[248] However, in the most recent overconsumption works, this increasingly becomes an argument about food.

Food seems to provide a particularly clear example of the personal dangers of overconsumption and the benefits of cutting down. Dauvergne, for example, explains his definition of 'wasteful consumption and overconsumption' purely in food terms, as 'consumption with no benefits for well-being, such as overeating until obese'.[249] Simms also uses food as an argument for reducing consumption, citing Second World War rationing as an example of how, by following Ministry of Food encouragement to 'Grow fit not fat on your war diet' and 'Cut out extras, cut out waste, don't eat more than you need', 'people did indeed become fitter and healthier and consumption of resources was drastically cut.'[250] This may be a rather simplified view of the effect of rationing on population health in 1939-45, since it disregards the effect of rationing not only in cutting the consumption of the better-off, but in actually improving the diets of the poor.[251] The idea that the chief diet-related problem in the UK in the 1930s was overeating is arguable, to say the least. However, this does reflect a common belief in the overconsumption literature about modern Western diets.

As Jackson comments in his 2009 report: 'When the American fridge freezer is already stuffed with overwhelming choice, a little extra might be considered a burden, particularly if you're tempted to eat it'.[252] This analogy for the benefits of cutting consumption clearly equates food, or modern, Western food, with the consumer consumption symbolised by the fridge. As an 'American fridge freezer', this is clearly supposed to be unnecessarily oversized and stuffed with food of which much will presumably be thrown away uneaten. While this is partly an argument about waste – there would be less risk of waste if the fridge were smaller and had less food in it – it's interesting that eating the food is presented here as the worst of the possible outcomes. Food consumption here has not only become consumer consumption, but among different types of overconsumption potentially the most serious kind. This is so much so that for some writers, overconsuming food can stand in for overconsuming anything else, as it is the major category. Clive Hamilton, for example, begins the chapter on overconsumption in his 2004 *Growth Fetish* with a discussion of obesity.[253]

The increasing prominence of food within overconsumption provides a theoretical backdrop for the specific arguments about food and its effect on climate change. It is a clear context for arguments that the solution to climate change lies in dietary changes by those who are judged to be overconsuming. So, for example, the drastic reductions in meat-eating called for by the Food Climate Action Network are included within general cuts in food consumption to what is clearly meant to be a basic needs level. The report recommends reducing alcohol, sweets and chocolate on the basis that they are 'unnecessary foods – they are not needed in our diet' and calls for us to monitor quantities as well as qualities of food consumption: 'do not eat more than you need to maintain a healthy body weight'.[254] They also make explicit the generally implicit shift in the overconsumption literature from a Marxist definition of items of basic needs to one set by external experts, as the report calls for a shift from a 'demand orientated' to a 'needs orientated' approach to food.[255] We might, after all, demand alcohol and chocolate, but the expert judge could point out that we do not physiologically need them. What is not explored in works like the Food Climate Action Network report, but is clearly important to the whole debate about food and climate change, is how the identification of overconsumption with food has changed who the overconsumers might be.

As discussed above, the groups in the frame for overdoing non-food, consumer consumption are those who have the material resources to afford it. For Galbraith, the urgent consumption problem of the 1950s may have been not so much middle-class consumption itself, but the increasing ability of working people to command sufficient resources to engage in it; yet this does not change the essentially middle-class identification of the behaviour. It is clear that more recent discussions of overconsumption in terms of consumer goods would also have reasonably wealthy people in mind. The idea of the harmful long-hours culture driven by consumer spending, for example, is only really applicable to those who are able to command high salaries for their long hours. If people have to work 60-hour weeks because of low pay, exhorting them to cut their hours and forgo their extra foreign holidays is not likely to have much effect. However, if problematic overconsumption is defined as 'eating until obese', the group responsible becomes different. Overconsumption is no longer about middle-class people competing for status through fancy cars and LCD TVs, it's about working-class people eating.

As we have seen, criticisms of working-class people's diets often take the form of encouraging them to eschew fast food in favour of healthier choices, so a defender of the conclusion that overconsumption is fundamentally a problem of working-class food consumption could argue that this is only highlighting the need for a change which would benefit working-class people themselves. The problem with this putative defence is that any argument which concludes that working-class people as a group are eating more than their share of the world's resources and are therefore the root of our ecological crisis is attacking them at a pretty existential level. After all, we need to eat to exist, not something which applies to earlier, higher-class definitions of overconsumption. It is also relevant that the identification of overconsumption as working-class food consumption has not developed in a vacuum. The idea that the consumption of the poor is problematic is not a new one, but one which has a long and inglorious history.

4

Malthus and the war on the fat and the poor

We shouldn't still have to talk about Malthus. For at least fifty years, writers on population have been dismissing him as outdated and irrelevant; that a nineteenth-century parson feared the poor would outbreed their resources and starve should not be worthy of any more than academic interest.[256] Unfortunately, we aren't there yet. Not only are Malthus' arguments about population still cited as worthy of serious consideration, some ways of thinking about overconsumption also seem to owe a considerable debt to a Malthusian world view.

Thomas Malthus (1766-1834) published the first edition of his *Essay on the Principle of Population as it Affects the Future Improvement of Society* in 1798, followed by a revised and extended second edition in 1803.[257] It sets out to show the existence of a natural Law of Population: that populations would grow until they outstripped their food supplies. This was, Malthus argued, because available food supplies increase arithmetically: you might be able to double your agricultural production in twenty-five years, but it would be unlikely that in the following twenty-five years you would be able to repeat the feat. However, in the same time, the population could double, and then double again, so your two-fold increase in food supply would be trying to feed a four-fold increase in population.[258] So, inevitably, people who could not afford the scarce supplies of food would starve, thus bringing the population back into line with its natural limits.

There are modern defences of Malthus' calculations. An essay in the *Optimum Population Trust* journal in 2008, for example, defended the basic correctness of Malthus' population equation, maintaining that while he had not been proved right yet, 'the truth of his proposition has long been apparent to anyone willing to see it, and the consequences of having ignored it will bear down on us in the immediate future'.[259] In the main, however, this outright defence of even the details of Malthus' arguments is a minority view; he is more often portrayed as both wrong and irrelevant. As the 1953 *Introduction to Malthus* pointed out, Malthus' approach to social policy was essentially that there should be no social policy, and that 'the social policies of industrialised countries are all alike, and Malthus could approve of none of them.'[260] Whether he can even be regarded as a precursor of the 'neo-Malthusian' school of population studies has been called into question, since it requires this defender of 'moral restraint' to have given rise to a movement which embraced contraception as the key method of population control (even if he might have approved of the neo-Malthusians' enthusiasm for forced sterilisation and eugenics).[261]

However, while the Law of Population is tacitly left to one side, the essential ideas of Malthus are clearly influential in modern population writing, whatever the writers take Malthus' central message to have been. Although Malthus lived long before the green movement, he is sometimes presented almost as an environmental activist. A 'bicentennial Malthusian essay' published in 1997, for example, did not try to defend Malthus' calculations on food production, but instead tried to adopt him as a climate change visionary. Malthus' importance, according to this view, lies not in the precise details of his Law of Population argument, but in his identification of what would go on to be the cause of climate change, ozone depletion and our other environmental ills: our 'indifference to limits'.[262]

This line of argument clearly relates Malthus' views on population to current debates about overconsumption, framing Malthus' Law of Population in terms of more widely accepted concepts of fundamental limits and carrying capacities. The idea that Malthus was a green could seem nothing more than a clever attempt by a fringe Malthus-enthusiast to fit his subject into a more relevant debate; a reaction to the irrelevance of the Law of Population to overconsumption and climate change. However, it is clear that even when Malthus is not referenced, Malthusian ideas fit naturally into much of the modern discussion.

The carrying capacity concept

As noted above,[263] population is a significant factor in the arguments about overconsumption. Even if overconsumption is considered as a measure of consumption per head, it seems obvious that consumption would increase along with the number of heads. While lifestyle and consumption choices may come into it, in these views ultimately the rate of consumption of resources, including food, across the globe is 'determined by the size of the human population'.[264] This is sometimes expressed as a direct, mathematical relationship, so, for example, it was calculated in 1986 that humans were already using 40% of the net primary product of photosynthesis, enabling others to draw the conclusion that a doubling of the human population would therefore mean that we would be using a clearly unsustainable 80%.[265]

A rather more sophisticated approach is that of the steady-state arguments. For the steady state, population is a key question, as an increasing population is seen as a major driver of the economic growth which causes environmental degradation. A zero-growth population would be a prerequisite for a steady-state economy, and this link at the theoretical level has also produced a degree of linkage between steady-state and population campaigning. The US-based Negative Population Growth calls for a no-growth, steady-state economy,[266] while the UK-based Population Matters is (at time of writing) listed by the Center for the Advancement of the Steady State Economy

(CASSE) as an organisation which has adopted a steady-state position.[267] In return, CASSE links to Negative Population Growth on its website, albeit with the caveated comment that 'Negative Population Growth is a national membership organization that tackles a very difficult issue head-on. We may not all agree on all their findings, but they have adopted a logical position called, "A No-Growth, Steady-State Economy Must Be Our Goal."'[268]

The connection between the steady-state theory and theories about the undesirability of population growth may provide a useful synergy for both. It works, but not because a no-economic growth position is inherent to the anti-population growth position. Malthus himself, after all, was very far from a proto-steady statist. In his later work, the *Principles of Political Economy*, he showed a distinct concern for economic growth in the absence of a growing population. The development of his theory of moral restraint was supposed to provide a stimulus to demand and therefore economic growth. His point was to make a distinction between 'prudential restraint', which he identified with restricting the number of children you have so as to fit your lifestyle to your income, and 'moral restraint', in which you strive to earn more money to be able to support the larger number of children which your industry enables you to have.[269] Malthus the good capitalist boy was convinced that economic growth was invariably beneficial. The connection of Malthusian arguments to the steady-state theory would probably have horrified him just as much as the use of his ideas to promote contraception. However the connection is a real one, as both the steady-state theory and the population arguments proceed from the same idea of fundamental limits – that the planet has a 'carrying capacity'.

The idea of a planetary carrying capacity is that it can only physically sustain a certain size of human population; that there is a fundamental limit beyond which human labour and technological development would not be able to extend the numbers which the world could support. As Pimental, a prolific writer on consumption and population puts it, 'When human numbers exceed the capacity of the world to sustain them, then a rapid deterioration of human existence will follow. As it does with all forms of life, nature ultimately will control human numbers.'[270] As discussed, this is a common concept in discussions of overconsumption,[271] but it's rarely acknowledged that the concept that nature limits us is also inherent to Malthus' arguments about population.

In his introduction to the second edition of the *Principles of Population*, Malthus argued that 'Through the animal and vegetable kingdoms Nature has scattered the seeds of life abroad with the most profuse and liberal hand; but has been comparatively sparing in the room and the nourishment necessary to rear them'.[272] While his Law of Population was about the comparative rates of increase in population and food production, the concept of a fixed, natural limit to production seems to underlie Malthus' thinking just as it does modern arguments about overconsumption. In discussing the potential to increase production in Britain in pace with population, for example, Malthus commented that 'In a few centuries, it would make every acre of land in the island like a garden'.[273] His calculations of the possible increase in production might place 'no limits whatever to the produce of the earth'[274] but an awareness of a fundamental limit was surprisingly close to the forefront of Malthus' work.

When overconsumption writers argue that the population has to be restrained to enable everyone to have a decent standard of living, they are echoing Malthus, sometimes explicitly, as Herman Daly, the founder of the steady-state school, did when he commented on that reasonable standard of living: 'We cannot precisely define "a good life", but most would agree with Malthus that it should be such as to permit one to have a glass of wine and a piece of meat with one's dinner'.[275] It is a measure of the development of the arguments about food and climate change that a comment written in 1996 should now sound so out of touch, since to achieve a steady state we are now told that a glass of wine and a piece of meat are precisely the things we should avoid having with our dinner, at least except on the most special of occasions. That Malthus as well as Herman Daly was concerned about carrying capacity is not just of historical interest, as these antecedents give a clue to the role of class in modern arguments about population.

Biological determinism

Malthus' theories were controversial as soon as they were written. Intended to refute an argument put forward by William Godwin that humanity could perfect itself through the application of human reason, the proliferation of *Principle of Population* editions was an attempt by Malthus to answer the criticisms he received. In the introduction to the second edition, he commented that 'I am aware that I have opened a door to many objections, and, probably, to much severity of criticism', although he found solace in the importance of his subject.[276] The third edition represented a distinct toning down of the first and second - 'I have endeavoured to soften some of the harshest conclusions of the first Essay',[277] Malthus said – and introduced the idea of restraint as an alternative to starvation for the poor. The central idea however remained, and so did the criticism. In 1830, for example, William Cobbett called the notion that people would always reproduce beyond the means of subsistence 'the infamous and really diabolical assertion of Malthus'.[278]

To Malthus and his supporters, this would appear as a form of shooting the messenger: he felt he was pointing out unpalatable but unavoidable natural truths, which would not become any less true as a result of critics railing against him. It's a form of argument which, over the last two centuries, has become part of the standard repertoire of right-wing social commentators. Claims that differences in average IQ test results between ethnic minority and white children are due to inherent differences in natural intelligence[279]; that all human behaviour is determined by our 'selfish genes'[280]; or indeed that women can't read maps[281] all proceed from this same standpoint: that inequalities in society are natural, rooted in biology, and can't be changed through political or social action. Indeed, for E O Wilson, whose *Sociobiology* launched a biological determinist school of studying human society when it was published in 1975, the 'ecological steady state' was itself dependent on advances in neurobiology, since until these were achieved 'a genetically accurate and hence completely fair code of ethics must also wait'.[282]

Malthus himself was at pains to present the unfortunate tendency of human populations to face mass starvation as a natural law, as a reality which affected all living things and which humans could also not avoid. As he said in the seventh edition: 'The race of plants and the race of animals shrink under this great restrictive law; and man cannot by any efforts of reason escape from it'.[283] However, what was naturally-ordained was that the poor should be the ones who starved. In the second edition, indeed, Malthus seems to be arguing that unemployment is a natural phenomenon, with the laid-off worker being rejected not only by the capitalist employer but by nature itself: 'At nature's mighty feast there is no cover for him. She tells him to begone.'[284] It was for this reason that Engels, writing at the end of the nineteenth century, called Malthus' work 'the most open declaration of war of the bourgeoisie on the proletariat'.[285] What Malthus presented as natural and immutable was, as so often turns out to be the case, an intensely class-driven argument.

It has been pointed out that biological determinist arguments have a tendency to emerge at particular political moments. Stephen Jay Gould comments in his criticism of IQ testing that the use of the IQ test as a measure of the hereditary intelligence of different ethnic groups was popularised in the US in the 1920s against the backdrop of anti-immigration measures and lynchings, while the three psychologists leading the charge reconsidered their position in the 1930s when 'PhDs walked depression breadlines and poverty could no longer be explained by innate stupidity'.[286] What Gould doesn't say is that the 1920s were also a decade of reaction against the upsurge of class struggle during the First World War and after. This had culminated in the 'Red Year' of 1919 and the Seattle general strike, which the mayor of Seattle called an 'attempted revolution'.[287] The idea that the poor were innately so was part of the vicious ruling-class reaction, just as the recurrence of the IQ arguments in 1969 was part of the reaction to 1968, and the publication of *The Bell Curve* in 1994 part of 'a new age of social meanness'.[288] In the same way that these upsurges of biological determinism are fully understandable only as part of the class struggle, so the background to Malthus' writing is as important to an understanding of the continuing meaning of the argument as the details of what he said.

A war against the poor in the nineteenth century...

The most obvious backdrop to a work about inescapable food shortages might be expected to be persistent food shortage and famine, but for the *Principle of Population*, this was not in fact the case. Malthus' lifetime saw significant increases in English food production, with grain production up by approaching 50% between 1760 and 1800.[289] Between 1750 and 1840 the population doubled, but production kept pace. In the 1830s, domestic grain production still covered 98% of domestic consumption.[290] This was the result of the eighteenth-century agricultural revolution, with advances in farming methods, development of new machines for threshing, harvesting and so on, and the introduction of new animal feeds all enabling food production to rise. It's this background which leads to the oft-repeated comment that Malthus was outdated even in his own lifetime, as industrialisation meant that the food crises which could hit communities reliant on subsistence farming could never again affect the Western world. We'll come back to this notion; for now it's simply worth noting that while Malthus was able to ignore food production soaring even as he wrote, he wasn't living in the past. If the *Principle of Population* was not reflecting an actual food crisis, it was nevertheless written out of a sense of an even more serious problem: increasing demands from the poor threatening the well-to-do.

The massive increases in agricultural production in the eighteenth and early nineteenth centuries were achieved on the back of a programme of enclosure which saw over six million acres of land, equal to about a quarter of the total cultivated land in England, turned from common, open or waste land into private fields.[291] Enclosure didn't happen to the same degree everywhere – it's been pointed out that in some areas by the mid-eighteenth century there was no common land to enclose – but coupled with the general inability of the rural poor, trying to support themselves from small pieces of land and some home-based manufacturing, to compete with industrial production and capitalist agriculture, it made it virtually impossible for them to be self-supporting. Poor people in the countryside were now reliant on wage labour to earn the money to feed themselves, rather than having direct control of their means of production. Flora Thompson, in her fictionalised autobiography about growing up in an Oxfordshire hamlet in the 1880s, described the difference enclosure made to the lives of the rural poor in the life story of the hamlet's oldest inhabitant, Sally, who 'could just remember… when [the hamlet] still stood in a wide expanse of open heath'.

> Country people had not been so poor when Sally was a girl, or their prospects so hopeless. Sally's father had kept a cow, geese, poultry, pigs and a donkey-cart to carry his produce to the market town. He could do this because he had commoners' rights and could turn his animals out to graze, and cut furze for firing and even turf to make a lawn for one of his customers. Her mother made butter, for themselves and to sell, baked their own bread, and made candles for lighting.

By contrast, Sally's husband worked all his life as a wage-labourer, 'for the cow, geese and other stock had long gone the way of the common'.[292]

The effect of enclosure was to drive up the requirements for poor relief, especially as agricultural employers relied on payments of poor relief to subsidise their low wages and maintain a permanent cheap labour reserve in the countryside which they could call on for temporary, seasonal labour. The amount paid nationally in poor rates increased from around £2 million a year in the 1780s to more than £4 million in 1803 and over £6 million in 1812.[293] Malthus' argument that the misery of the poor was both natural and all their own fault therefore gave ideological cover for the enormous theft of rural land by the bourgeoisie to which the enclosures amounted, justifying the effects of a process of which he must have been well aware. More than this, Malthus' Law of Population also provided ammunition for those who were getting tired of covering for their own and their competitors' disinclination to pay their labourers a living wage. His argument that the poor should not be enabled to reproduce if they couldn't afford it certainly helped the move towards the 1834 Poor Law, which limited the amounts paid in poor relief by denying any support except the workhouse to those deemed capable of working. Conditions in the workhouses were deliberately designed to be unpleasant, based on a much-admired workhouse in Southwell, where conditions were 'as disagreeable as was consistent with health', and importantly, husbands and wives admitted to the workhouses were deliberately kept apart so that they couldn't have any more children.[294] It is not for nothing that the 1834 Poor Law has been dubbed as Malthus' gift to the bourgeoisie.[295] As an 'early and intimate friend' of Malthus wrote shortly after his death in 1834:

> No intelligent and well-educated person can have observed what has been passing in the civil economy of this country for the last forty years without being convinced that a great change has been gradually wrought into the public mind respecting the poor laws and their administration, and that the works of Mr Malthus have been exceedingly influential in bringing it about.[296]

Saying that the rural poor were dependent on poor relief was therefore a convenient way of concealing the lack of the means of employment by confusing them with the means of subsistence. As both Marx and Engels pointed out, the problem for the rural poor was not that there was an absolute lack of resources to feed them, but that they were deprived by enclosure and capitalist agriculture generally of the means to support themselves, leaving them reliant on low-paid, intermittent wage labour for subsistence.[297] This was a point made vividly by William Cobbett, writing in 1830 about the poor of the Avon valley, when he commented that the population of the area seemed to have fallen compared to earlier periods, but that the people who remained seemed to be no less poor, despite the fact that the land could presumably have supported many more. He asked 'where, then, is their natural tendency to increase beyond the means of sustenance for them? Beyond, indeed, the means of that sustenance with which a system like this will leave them.'[298] Malthus conflated the effect of capitalist relations of production with carrying capacity by constructing the former as part of the latter:

at least as far as the first edition of the *Principle of Population* was concerned, choices made by employers to lay off workers were an expression of the natural limits of possible support for the poor. Capitalism's need for a reserve of cheap, seasonal agricultural labour was for Malthus an aspect of the inescapable iron law of nature, which no human society could escape.

...and in the twentieth century

This tendency to conflate the effects of economic and social structure with absolute natural limits seems to have been inherited by some of the most infamous modern writing about population. Malthus may have been inspired by the growing requirements for poor relief at the end of the eighteenth century. In the same way, a sense of privilege being besieged by growing hordes of poor people, reminiscent of Malthus' writings, pervades notorious works like Paul Ehrlich's 1969 *The Population Bomb*. Here he explained how he was on holiday with his family in Delhi when he first reached the conclusion that the world was so overpopulated that a billion people would starve by 1983. It's a passage which it seems almost obligatory to quote when discussing population, but it does need to be read to appreciate the full classist and racist horror of the white, middle-class academic facing the poor, numerous and brown: 'The streets seemed alive with people. People eating, people washing, people sleeping, people visiting, arguing and screaming... People, people, people, people. As we moved through the mob, hand horn squawking, the dust, noise, heat and cooking fires gave the scene a hellish aspect. Would we ever get to our hotel? All three of us were, frankly, frightened.' He concluded, 'since that night, I've known the *feel* of overpopulation'.[299]

Ehrlich belonged to an apocalyptic school of population thought which is perhaps less fashionable now than it was forty years ago, although this style of argument has not entirely disappeared. The obvious point that the number of people in a crowded city street is in no way an expression of the size of the population of a country (or even of a city), let alone evidence of said population's relationship to its productive capacity, has been made many times, but it is still possible to find arguments about overpopulation which refer directly to Ehrlich's Road to Delhi experience. Lindsey Grant, for example, writing in 1996, claimed that the evidence for catastrophic overpopulation was there for anyone to see: 'The numbers are visible enough. Try a visit to Calcutta or New York.'[300] The choice of these particular cities for evidence of the population problem shows that the racist assumptions underlying Ehrlich's writing are unfortunately still there in more recent population arguments. Indeed, there is a clear synthesis between anti-immigration arguments, particularly in the US, and the idea that the population of poor, non-white countries is out of control.

However, even leaving aside this racist, explicitly right-wing line of argument, in more respectable discussions of population the lines of descent from Ehrlich and Malthus can be clear. Given that the argument is supposed to be about the ability to feed the entire population, it's notable how frequently concerns about population appear as concerns about literal space. When this takes the form of discussions of congestion, pressure on public spending and so on, the most obvious sources are wider discussions of immigration, which routinely use the arguments that migrants will push up housing costs and cause longer traffic jams in major cities, as support for restricting numbers. Alongside this however are concerns not about the addition of people to already crowded spaces but the incursion of people into previously unpopulated areas. It hardly seems the most serious of the concerns in the population debates, but the idea that national parks will become too heavily used turns up with surprising frequency.[301] Indeed, Virginia Abernethy in her introduction to Ester Boserup's famous work *The Conditions of Agricultural Growth* criticised Boserup, and by extension a general approach to population, for ignoring what she felt was the deeply important national parks issue: 'the value that some people place on wilderness, space, mobility and other intangibles that lose out to crowding and population growth'.[302]

The point is that this is not only a continued use of a visceral idea of overpopulation unrelated to population statistics, but also that it identifies a key effect of overpopulation as the incursion of additional people, presumably poor, possibly non-white, into previously privileged spaces. The references to national parks in the US is indicative, as these are spaces once the preserve of those with relatively rare time and transport to reach them but now accessible to the wider population, with the implied concomitant that they are now spoiling them for the more discerning. This is, of course, the same basis on which consumption becomes problematic overconsumption: when it's consumption by the masses and not by the privileged few.

The elements of racism and overtly right-wing sentiments which appear in many expressions of concern about population have made it an uncomfortable issue for much of the environmental movement, but it is not easy to get away from it. How to treat the population issue remains a controversy within the Green Party of England and Wales, for example. When the Green Party leader, Natalie Bennett, wrote a letter to *The Guardian* in July 2013 to take issue with the 'vicious rhetoric on immigration' poisoning British political debate,[303] three Green Party members wrote in reply to take her to task on the basis that immigration was a key driver of population growth and should therefore be opposed by any environmentalist party.[304]

The argument that immigration is an environmental issue has also been raised in a US context and it has differing explanations depending on who is proposing it. In the US, the argument seems to be mostly that immigration from a Third World to a First World country would usually raise the living standards of the immigrants, therefore meaning that their *per capita* emissions would be higher. They might even visit a national park or two while they were at it. Others indeed take the view that not all parts of the globe have the same carrying capacity, so that, for example, an increase in potential visitors to Yellowstone would be a problem, or that it might be better for the environment if Poles remained in Poland where there is more space per person rather than moved to the UK.

Sandy Irvine, one of the authors of *The Guardian* letter, says that the Green Party leadership do not want to engage with the population question, to the extent of preventing Population Matters (of which he is a trustee) from putting a leaflet into the party publication, *Green World*.[305] He is determined that a concern about population is neither right- nor left-wing and that environmental campaigners are making a mistake in ceding the ground on immigration to the likes of the UK Independence Party (UKIP). We were speaking a week after the 2014 European elections in which UKIP gained the highest vote across the UK, so this was a topical concern. For Sandy, an environmental position which leaves out population is omitting a third of the areas it should be covering. Greens, he says, have to consider the technological causes of climate change, consumption per head and the numbers of heads doing that consuming; anything else simply ignores huge aspects of the problem, including the hard truths that shifting to a sustainable society with this many people will have to involve significant sacrifice of living standards in the West.

His own understanding of population as a serious issue came, like Paul Ehrlich's, from personal experience of overcrowding. Growing up on a council estate in Huddersfield but with family in the Shetlands, as a child he was always very struck by the difference between the estate and the family croft, 'the difference between a crowded place and a very uncrowded place'. It was this experience of the contrast which first made him think about issues of space, density and privacy, an experience which he brought with him when he went on to study town-planning in Newcastle. This was 1968, and the focus of the course was on coping with growth. The students were expected to design big new towns, which would have flattened local villages like Stannington, to the north of the city. While part of this was a drive to provide better housing, Sandy saw it as also, in part, a response to growing human numbers which meant that uncrowded, green spaces had to be sacrificed. At the same time, the famous 'Earthrise' photograph (the picture taken by the Apollo 8 crew in 1968 of the Earth rising from the Moon) brought home to him that we live on a finite planet: human numbers could not simply go on increasing.[306] This realisation took him politically from the International Socialists to the Ecology Party (later the Green Party) and the Optimum Population Trust (later Population Matters), but his story also emphasises how the population issue and discussions about overconsumption are interconnected.

While the political orientation of many of those proposing population as the key environmental problem might seem to set the population arguments aside from much of the environmental movement's discussions of overconsumption, it appears that the underlying views of the problem are more similar than it might be comfortable to admit. For both, it is the existence and consumption of the poor which is the problem, whether it is the food choices of working-class people in the First World, or the reproductive choices of the poor in the Third World. Approaches to these problematic people can be more or less sympathetic: some of those concerned about overcrowding in US national parks give the impression that they would prefer access to them to be restricted only to the privileged, while Sandy Irvine, for example, clearly came to his position through his own experience of growing up in a working-class area. It is worth noting, however, that whether they are to be condemned or not, the effect of the population arguments, and their off-shoot that immigration is an environmental issue in particular, is that they cast as the world's number one climate criminals not the chief executives of oil companies (to take an example at random) but recent immigrants to the West. The same people, in other words, who are singled out as the sharp end of the obesity explosion, destroying the world with their fast food habits. The overconsumption and the population arguments come together here to identify the problem as one of growth of the working class in the developed world.

The idea of out-of-control population growth clearly carries with it the threat of redistribution from the rich to the ever more numerous poor. As Susan George put it: 'Certainly we are afraid – afraid that increasing numbers in the Third World will one day demand from us their due and lower our own standard of living; fearful that the pressures of population may finally demonstrate that the "only solution is revolution"'.[307] This is the same fear that Engels invoked in 1892, when he commented that the idea of allowing the surplus population to starve was 'simple enough, provided the surplus population perceives its own superfluousness and takes kindly to starvation… the workers have taken it into their heads that they, with their busy hands, are the necessary and the rich capitalists, who do nothing, the surplus population'.[308] The overconsumption arguments are more likely than discussions about population to embrace the idea that the just way to deal with this situation would be to equalise consumption across the world, rather than to take punitive measures to prevent the First World being deemed the surplus population. However, even with this approach it is difficult to see this view of the cause of climate change and environmental destruction as positive for the working class.

Technology versus carrying capacity: arguments against Malthus

The statement that on a finite planet there is a fixed, natural limit to production may seem an obvious one, but its practical implications can be disputed. While it may be that, logically, a limit to the productive capacity of the earth must exist at some point, it is by no means so certain that we are likely to be anywhere near that limit. Against the Malthusian view of a fixed, proximate carrying capacity, there is therefore the argument that human ingenuity and technological developments will be able to increase production for all practical purposes indefinitely. This was argued by Engels in his *Outlines of a Critique of Political Economy* in 1844, only a decade after the last edition of Malthus' *Principle of Population*, when he stated that human labour could always extend what might appear to be natural limits to production: 'the productivity of the land can be infinitely increased by the application of capital, labour and science'.[309]

This can't be dismissed as simply an expression of a nineteenth-century belief in 'progress' as the idea was developed in the 1960s by Ester Boserup into a full-scale theory of agricultural innovation. Boserup argued that population pressure, far from being a threat to human societies, was actually the key driver of technological development. She took issue with Malthus' view of static or only slowly-increasing production; looking at pre-industrial societies, she found that they were able to produce more food as their population increased through developments in agricultural techniques. The Malthusian way of looking at the relationship between food production and population was in fact the wrong way round: Malthus saw population size determined by food availability, so the population would increase if more intensive farming methods made more food available, whereas lack of food would reduce the population through famine. On the contrary, Boserup argued that the population increase came first: it caused the intensification of agriculture and greater food production, not the other way round.[310]

Boserup's theories were influential for thinking about population and development. In 1984, for example, the World Bank gave a decidedly Boserupian summation of the problem of population for developing countries: 'The difficulties caused by rapid population growth are not primarily due to finite natural resources, at least not for the world as a whole',[311] but to the lack of capital to invest in technologies to expand production. Boserup had developed her theory on the basis of pre-industrial, non-capitalist societies and argued herself that it would be difficult for poor rural communities to industrialise, even if that would increase the amount of food they could produce.[312] This general school of thought about the effect of technology on production also managed to win over some Malthusians, who were reduced to arguing that Malthus was right, but only up to the Industrial Revolution. The French historian Emmanuel Le Roy Ladurie, for example, believed absolutely that the peasants he studied in the sixteenth- and seventeenth-century Languedoc were subject to the 'Malthusian curse', as were 'certain peoples in today's [1966] Third World', but in general, industrialised agriculture had made Malthus irrelevant to most modern societies. He was 'a clear-headed theoretician of traditional societies, but he was a prophet of the past; he was born too late in a world too new'.[313]

The fairly widespread acceptance of the idea that technology would trump limits to production means that it can't be categorised as a purely left-wing one. Indeed, it would be difficult to claim as purely socialist any argument adopted by the World Bank. However, just as the concept of a relevant and proximate carrying capacity can be linked to right-wing arguments about the superfluity of much of the working class, so arguments which maintain the value of human labour over natural constraints have tended to be used as a rebuttal by the left. This is sometimes on ideological grounds. Lenin, for example, saw neo-Malthusianism as the pessimism of the petty bourgeoisie, compared to the determination of the proletariat not to have to despair of the future. The workers, in contrast to the petty bourgeoisie, were 'fighting better than our fathers did. Our children will fight better than we do, and they will be victorious.'[314] More often, however, the argument appears as one of empirical fact rather than one of political approach: conveniently for those of us who find the implications of the carrying capacity argument disturbing, it happens to be factually incorrect. So, a recent pamphlet on *Marxism and Ecology* cites a UN FAO report that cropland could be doubled, and that in the Guinea savannah only 10% of land suitable for farming is actually being cropped.[315]

These are comforting statistics, but it isn't as easy as that to dispose of the idea of carrying capacity. The first problem is that, while Malthus may have used the image of the entire land surface being taken up eventually by farming, carrying capacity as it has developed is not mainly or even primarily about land use. It is unusual to find arguments that there are physically no more pieces of land which could be turned over to agriculture; indeed, figures for the amount of land which could potentially be cultivated but isn't tend to be both contradictory and surprisingly hard to come by. It is perhaps for this reason that overconsumption writers like Goodland who want to put their arguments on a mathematical basis adopt measures like the percentage of the primary product of photosynthesis in use rather than estimates of land surface availability.[316] In any case, a measure of how much land is not currently being used for farming would not be a measure of how much could be brought into use. With deforestation a major cause of greenhouse gas emissions, the climate effects of clearing land are an integral part of the modern idea of carrying capacity, as are water resources, the shortage of which, exacerbated by climate change, may be as effective a brake on agricultural development as land availability *per se*.

The point is that carrying capacity has to be understood not just in terms of how much production can be physically wrung from the planet, but also of the effect that a certain size of population has on the biosphere. In this view, population size and production go hand in hand, although this is of course arguable. Climate change here is itself evidence that we have exceeded the planet's carrying capacity. The effects of trying to fit too many people on to the planet are not first evident in starvation, as Malthus thought they would be, or certainly not in the developed world, but although we aren't yet suffering the personal consequences, the size of our population means that we are treading too heavily upon the earth. In this argument, it might be possible for us to continue to produce enough food to feed everyone for some time, but the climate effects of such an attempt would mean that it would be too destructive to the environment for us to continue to try to do so. Whether or not the argument goes on to conclude that the remedy is to reduce the numbers of people, or for everyone in the West to modify their living standards, it's clear that simply thinking about the amount of land which is not currently being farmed is insufficient to answer this argument.

Nor will a recourse to human ingenuity do any longer as a counter-argument to carrying capacity. The industrialisation of agriculture which for writers like Ladurie made Malthusianism irrelevant is of course reliant on fossil fuels, which are not only a major cause of climate change but subject themselves to fundamental resource constraints. It is likely that we are at, or nearly at, 'peak oil', the point at which depletion of global oil reserves means that output begins to decline, and it's clear that this will have serious effects on food production. In 2008, for example, Chatham House's report on the future of food considered various different scenarios for future food production in which significant differences in food availability depended almost wholly on the price of oil. In the worst option, as food production falls in the face of fundamental resource constraints, 'the view spreads that peak oil has arrived', making clear that for Chatham House, the resource which determines carrying capacity is not land, but oil.[317]

Oil is clearly subject to relevant and proximate constraints; even if new sources of oil are found to avoid peak oil, the climate requires that we don't attempt to use them. Attempts to use oil sources which have previously been thought too difficult to be worth the expense of extraction, like the Canadian tar sands, will cause immense environmental damage even before the oil extracted is used. The obvious counterargument here is that human ingenuity does not have to stop at fossil fuels: we could develop new technologies which enable us to develop industrial-scale agriculture without the currently-attendant greenhouse gas emissions. However, the problem from the perspective of much of the green movement is that attempts to expand food production through technological development do not have a good track record. We can no longer justify the confidence of the 1840s, or indeed the 1960s, that capitalism will always be able to work and invent its way out of food production limits.

The problem with the green revolution

The backdrop to the confidence of writers like Ladurie in all-conquering modern technology was the green revolution. This was a development programme for higher yield crops, begun in the 1940s when the Rockefeller Institute started funding research into wheat strains, which was beginning to see results in the 1960s as the first higher yield strains developed were being planted. The results were spectacular: in Asia, for example, green revolution rice strains increased yield by 30% more than the increase in population between 1963 and 1993 across the continent as a whole.[318] It's easy to see how, from the standpoint of the 1960s and 1970s, the green revolution could have seemed to have solved the contemporary problem of low food production, and averted the population apocalypse being predicted by people like Paul Ehrlich. With this as an example, the idea that future food problems could always be averted by new research and development was an obvious one, but it's not so obvious now.

The green revolution is controversial because it didn't just create higher-yield varieties of staple crops, but higher-yield varieties of staple crops which were reliant on large-scale, mechanistic farming methods. While overall yields certainly increased, this wasn't always a benefit to the poorest farmers, who were unable to compete with marketised agricultural concerns and were often driven off the land altogether. There is a clear comparison between the effects of the green revolution and the effects of enclosure and capitalist agricultural production in Britain in the eighteenth century, which as discussed was the background for Malthus' thinking on population. The high yields of green revolution crops were gained under the assumption that intensive commercial agriculture is the only way to produce large amounts of food, despite alternative evidence that peasant agricultural techniques can be both more productive and more efficient.[319]

It could be argued that these aspects of the green revolution do not invalidate the advances it represents in terms of food production. Fred Pearce, for example, takes on scepticism about the benefits of the green revolution in particularly forthright style:

> Many people I know regard the green revolution as a disaster. They say it has tied billions of the world's peasants to a marketised, globalised, mechanised, energy-guzzling, climate-

> warming, biodiversity-destroying way of feeding the world. I see their point. And it might have been done differently. But would they prefer billions starving?[320]

This holds out the hope that future developments of food production technology could indeed 'be done differently'; that we don't have to abandon belief in human ingenuity simply because the green revolution turned out to be a two-edged sword. However, it relies on seeing the environmental and social consequences of the green revolution as extrinsic to the project, when a more appropriate view may be to see them as an indivisible part of advances in capitalist agriculture.

The introduction of the limited numbers of high-yield strains has had the effect of destroying diversity, and therefore increasing vulnerability to disease among the remaining strains of the crops. In India, for example, one result of the green revolution has been forty new insect pests on rice and twelve new rice diseases.[321] The new crops tend to require large amounts of pesticide, and have also proved to be much thirstier, and therefore require more water, than older strains. Both of these factors make the crops difficult to justify on sustainability grounds, as both water and oil resources become scarcer as a result of peak oil and climate change. Again, it could be done differently next time, although the development of genetically-modified (GM) crops along the same lines, with similar problems of corporations expropriating crop strains, reliance on expensive fossil-fuel fertilisers and pesticides and destruction of diversity, suggests that these issues are rather more intrinsic to this type of development than defenders of the green revolution would like to think. In addition, however, it is also possible that the expansion and improvement of agriculture through technology itself is subject to limits.

The green revolution worked not by simply supplying existing cultivators with higher yield crops, but through the appropriation of peasant holdings to create neoliberal agro-export regimes. Just as with enclosure in eighteenth-century Britain, this did lead to increased production, but it is not an increase which can be repeated indefinitely, even if we were prepared to put up with the social costs. The production gains of the green revolution were pretty much exhausted by the 1980s, and even if a way were found of increasing yields without requiring more fossil fuel or water resources, the initial gains might not be repeatable.[322] This of course is not to argue that it would never be possible, under any system, to improve agricultural production. However, the green revolution should be seen as not so much the application of superior technology as an exercise in bringing agricultural land into capitalist production; an exercise which can only happen once.

Capitalist exploitation of natural resources works through the availability of new areas, outside those currently used for production, to expand into: continued profit requires ever more new resources to exploit. The green revolution was an expression of this as it worked by bringing land which had not previously been part of capitalist agriculture into the global market system. If the increase in productivity which it achieved were to be repeated in the same way, it would require there to be more land currently outside the capitalist system, which could be brought into it to be exploited. This availability, the existence of non-capitalist areas for capitalism to expand into, obviously cannot be infinite, whatever we might think of human ingenuity's ability to invent new technologies. As Engels pointed out, despite his optimism about the ability of technology to improve production, within capitalism, the hope of feeding the world's population, including the burgeoning population of industrial cities, came ultimately from the fact that there were large areas of the world which had not yet been exploited. Writing in 1865 of the untapped resources in places like the western USA, he commented that 'If all these regions have been ploughed up and after that shortage sets in, then will be the time to say *caveat consules*'.[323] The Latin translates as 'let the consuls (rulers of republican Rome) beware', an indication of the revolutionary possibilities of an era of profound food shortages and a far cry from Malthus' conclusion that if the poor are starving, they should resign themselves to doing it quietly.

It would be an easy response to the arguments which make ordinary people essentially responsible for overconsumption and therefore climate change to be able to argue that human labour and ingenuity will always resolve the problem, but it is difficult to make this case with as much confidence as Boserup did in 1965. The idea that humans can always overcome natural limits in fact appears to be a product of a type of alienated thinking which sees the natural world as separate from, and therefore potentially exploited and dominated by, human society. It is a similar view of nature, incidentally, which sees it essentially as untouched wilderness, without the despoiling presence of poor, working-class humans: nature as something the wealthy go to visit, rather than the world of which human societies are part. We have now seen in climate change the effects of this view of the environment as separate from us, and the onward march of progress is no longer an attractive argument.

Neither Malthus nor capitalist production: a way through the impasse

If we can't argue any longer that technological development within capitalism has and will continue to make Malthus irrelevant, does this then mean that we have to accept Malthusian-influenced notions of a carrying capacity? This would still leave us room to dispute ideas about how close we are to the limits of that capacity, and what we might do about it. Simon Fairlie, for example, defended meat-eating in 2010 on the basis that limited livestock farming could be a strategy for avoiding the otherwise inevitable drift of human populations towards outgrowing their resources, essentially by acting as hoofed miners' canaries for overconsumption.[324] Whatever the conclusions might be about the precise composition of our ideal diets, it could seem at this point that, without the trump card of the Industrial Revolution, we are stuck with Malthus' conclusion that at some point, whether imminent now or not, there will be too many people in the world and the only solutions will be the drastic reduction of *per capita* consumption, or reduction of population. Fortunately though, we aren't.

The problem with the notion that human work or human inventiveness will overcome natural limits is that it assumes that there are natural limits to which, if humans are too idle or too conservative, they will be subject. This is a trap which Boserup largely avoided, since she was working on peasant communities mostly using non-industrial methods of improving their harvests, but which writers like Ladurie fell into wholesale in concluding that only industrialisation invalidates Malthus. The general idea here is that pre-industrial societies were subject to recurrent absolute dearths of food sufficient for their populations because their agricultural production methods were so inefficient, and it is only the advances of industrial production which has enabled our food production to keep up with population. There is a long list of catastrophic famines in the medieval and early modern periods to substantiate the idea that human societies before the Industrial Revolution have always been no more than one step ahead of starvation. This however relies on a key assumption about the causes of those famines which is not borne out by the evidence. A review of the history of famine shows that it is not that Malthus was wrong about food and population in his own time, after industrialisation. It is that he was never right.

A (very) short history of famine: all about class

The classic view of pre-industrial agriculture is of a system at the mercy of drought, storm, disease and other factors which could cause harvest failures. If in good years all the food produced was required to feed the population, then the surplus saved against bad harvests would be very small. Hence instances of famine caused by harvest failures can be taken as evidence of an underlying Malthusian problem: that the population was pressing on the limits of production. This attribution of single events to an underlying Malthusian cause is extended by some from famines to disease, as Ladurie, for example, blamed even the Black Death on overpopulation, calling it the 'holocaust of the undernourished'.[325]

This doesn't exactly hold up historically: while some recurrences of the plague did seem to affect specific groups, like the 1361 outbreak which seemed especially to hit young adults, one of the many horrifying things about the Black Death for contemporaries was the way in which it wasn't restricted to the poor or already ill. The high death rates among well-fed groups like the clergy are indicative of this. Ladurie seems to have been thinking of the Black Death as a Malthusian consequence of overpopulation in much the same sense that Malthus himself thought that an aware, teleological Nature would tell the unemployed farm labourer to 'begone'. The medieval world was overpopulated and hence asking for it. It is possible to speculate that writers like Ladurie had to introduce disease as a Malthusian mechanism because famine is actually an ineffective reducer of population. Out of all recorded famines, only the Chinese Great Leap Forward may have come close to killing more people than the 1919 flu pandemic, while in sub-Saharan Africa, the bigger killer is now not famine but Aids.[326]

The Industrial Revolution is of course indivisible from capitalism, and underlying arguments that pre-industrial

Europe was in a Malthusian situation is the idea that only the market economy can provide the stimulus both for technological development and to make peasants work harder to produce more food. There are medieval precedents for this view, but it fails to understand the real constraints on peasant production. It is probably true that pre-modern European agriculture was under-productive, but laziness on the part of individual peasants who would rather starve than work is unlikely to have been the cause.

One clear instance of underproduction, for instance, was the reliance through much of late medieval Europe on one cereal crop per year. This exacerbated the risk of catastrophic harvest failure and clearly limited the amount of food which could be produced, whereas even with pre-industrial agricultural techniques it should have been possible in many places to get two crops in a year. It has been suggested that the reason for this failure to maximise food production lay in the structure of late medieval society, in which landlords' demands on their peasants acted as a positive disincentive to produce surplus grain, as the lion's share of it would simply be appropriated. The exploitation of the peasantry also meant that it would have been very difficult for peasants to raise the capital, or indeed to devote the physical resources, to grow a second crop.[327]

This is an illustration of the fact that production before the Industrial Revolution was determined, just as much as production after it, by the structure of the society doing the producing. The tendency of the Malthusian arguments is to regard pre-industrial society as one in which individual peasants confronted natural food production with no social structure intervening, but this is of course inaccurate. Famine is not a natural phenomenon, unmediated by the mode of production operating in the society in which it occurs; it is a social phenomenon. Indeed, an examination of the history of famine reveals that it has always been all about class.

It's easy to assume that famines represent the absolute lack of food in a particular area: not necessarily that there is literally no food at all, but that the amount of food available divided by the population ends up with shares which are insufficient to sustain life. However, it seems that this situation is actually extremely rare. It is possible that an Indian famine in 1344-45 may have involved an absolute lack of food, since its effects seem to have been felt so far up the social scale that it affected a king, but as this same king is also described as organising relief efforts, even this example is disputable.[328] Even if it is accepted as an example of a famine in which no-one was able to get enough to eat, its rarity shows that the correct understanding of what a famine usually entails is that it is not an absolute dearth of food, but a situation in which some people go hungry while others eat.

The best known, although by no means the only example that demonstrates this point, is the Irish Potato Famine of 1846-51, in which at least one million people died. In reaction to Malthusian-style thinking about famines as demonstrations of populations which have outgrown their food supplies, some have argued that famines do not have to entail a food production problem at all. Amartya Sen thought that this was the case, for example, with the Bengal famine of 1943-44, attributing it to hoarding done to raise prices and to the diversion of grain from Bengal to elsewhere in the British Empire for the war effort.[329] His point was that starvation represents a failure not of the production of food but of people's entitlement to it: 'The law stands between food availability and food entitlement. Starvation deaths can reflect legality with a vengeance.'[330] Sen's interpretation of the causes of the Bengal famine has been criticised, as despite attempts by British officials at the time to blame the problems on hoarding by merchants, there is evidence of genuine food shortages arising from harvest failures.[331] These would not, however, have led to a famine on their own; ultimately, the poor in Bengal would not have starved in 1943 had the war not put them far down the list of British imperial priorities. The Irish

Potato Famine demonstrates a similar dialectic between the proximate causes of famine and the structural reasons for famine deaths.

There certainly was a fundamental food shortage in Ireland, as the potato blight fungus destroyed much of the potato crop in 1845, 1846 and 1848. This was however only part of the story, as the British government failed to organise anywhere near sufficient famine relief. They spent £8 million on dealing with the famine, much of that on protecting landlords' property from their starving tenants, compared to £70 million spent on the Crimean War. Not only that, they also refused to take the obvious initial step once the potato crop failed and continued to allow grain exports from Ireland. In 1846-47, Irish people who objected to seeing grain which could save their lives being sold abroad for profit started taking matters into their own hands, with agrarian secret societies threatening farmers who sold their grain for export and even shooting the horses used to take the grain to market.[332] As some critical voices pointed out at the time, the million Irish who died, died from the application of market forces as much as from potato blight.

The diehard Malthusian response to this would be that even if the effects of the famine were exacerbated rather than ameliorated by the British government, the Irish were put in the position of requiring famine relief in the first place because the population had outgrown its productive capacity. The population of Ireland increased rapidly in the eighteenth and early nineteenth centuries, and potato cultivation was a way of feeding large numbers of people on little land. The Malthusian view of the famine would be that the Irish, improvidently unaware of the dangers of reliance on a monoculture, created their own tragedy. However, it was not the size of the population which drove the increasing Irish reliance on potatoes.

Potato cultivation had been an important part of agriculture in Ireland since the seventeenth century, but for most it was a supplement to, rather than replacement of, cereal farming; a backup in case the grain harvest failed. More and more of the rural poor became entirely reliant on potatoes for subsistence during the course of the eighteenth century, not because they were having too many children to be able to support them any other way, but because of the effects of British imperialism. Large-scale agriculture in Ireland was increasingly aimed at grain production for export to England, while the poor were forced on to smaller plots in more marginal land.[333] This was a process that continued as a result of the famine itself, as Irish landlords used the famine and non-payment of rents to evict about half a million people, in what John Newsinger rightly calls 'one of the most terrible acts of class war in modern European history'.[334] The failure of the potato crop made a crisis out of a situation which had arisen not because of the existence of the poor in Ireland but because of the expropriations of the landlords and the British. The famine was the creation of the system, not of the size of the population.

The importance of the mode of production

The concept of carrying capacity posits a limit to food production which is outside any economic or social structure. It says that regardless of how we organise ourselves to produce food, there is an absolute limit beyond which we will not be able to go. For Malthus, and for enthusiastic Malthusians like Ladurie, the proximity of this absolute limit is illustrated by the history of famines in non-industrialised societies. This is a history which it could seem we are facing again, unless we make significant cuts in food consumption, a discussion which appears to focus on consumption by the comparatively poor, rather than by the rich. However, what the history of famine indicates is not fundamental limits but the effects of exploitation. Even looking at pre-capitalist societies does not allow us to think that we are looking at the effects of natural limits to food production, rather than the effects of the mode of production.

Marx pointed out that what Malthus was seeing with his Law of Population was not a natural law but the process by which capitalism produces a surplus population. Capital is divided into constant and variable capital: the money used to buy machinery, property and so on versus the money available to pay for labour. As capital's demand for labour is determined by the variable component of capital, it falls as more capital is accumulated. In effect, workers produce the means by which they are turned into a surplus population, as it is the surplus value which their work generates which allows capital accumulation in the first place. The existence of a surplus population, while bad for workers, is very useful for capitalist businesses, as it provides a pool of labour from which they can recruit quickly if they need to expand production, and the means to keep labour costs down through the constant threat of unemployment for any workers who might think of demanding better pay and conditions. It is an intrinsic part of how capitalism works, but capitalism is not a natural phenomenon.

Malthus' argument works by treating the effects of capitalism as if they are both natural and universal. For him, the tendency of capitalism to produce a surplus population was a fundamental resource constraint. The Chatham House report on the future of food treats our current dependence on oil in a similar way, as a universal truth about food production which cannot be changed. But if we are considering whether we are now up against the carrying capacity of the planet, we have to be able to look outside the constraints of the current system. It is part of the atomising effect of capitalism to consider food consumption as a matter of individuals with problematic behaviours. If we really want to understand how the food production system is contributing to climate change, and whether we can all continue to be fed, we have to go beyond individual actions to consider the actions of the food production system within capitalism. It may come as a shock, but we may not have the most efficient, least wasteful way possible of producing food and living within any fundamental limits that might exist.

5

Waste and the limits of capitalism

The concept of natural limits to food production is in many ways a simple one. It has become part of the assumed framework for much green thinking about food and climate change precisely because it is based in what seems to be such an obvious, common-sense reality: we live on a finite planet, and we only have the one. This is of course true, but as discussed in the previous chapter, the real question is not whether or not there is an eventual natural limit to production, but whether we are anywhere near that limit. Assessing this is not a simple calculation. While there are a number of expert attempts to carry it out, they have a tendency to disagree with each other: according to the IMF, for example, food demand is likely to outstrip supply by 2080, whereas other, more confident, predictions hold that food supplies then would still be sufficient even if the global population were to double.[335]

For Malthus, as we have seen, it was all about land use and space. His vision of the effects of unchecked population growth was of more and more land in Britain being cultivated, until the entire island was taken up with gardens. This leads on to modern neo-Malthusian concerns about crowding in 'wild' spaces like US national parks, and perhaps to a more general worry about physical lack of space on a future planet with a larger population. The disappearance of large tracts, if not absolutely all, of the planet's landmasses under a continuous urban sprawl is a fairly common feature of futuristic fiction,[336] while Malthus' ultimate nightmare was indirectly dramatised in George Lucas' *Star Wars* prequels, in which the planet Coruscant is completely covered in one big city. However, this isn't really what coming up against natural limits would look like. Unlike Malthus, we understand that the planet's complex ecosystem would not allow total cultivation, or indeed urbanisation: even if we wanted to cover the entire land surface with a city, or with gardens, the environmental strain would be simply too great. The concept of natural limits is therefore a measure not of physically how many plants could fit onto the surface of the planet, but how much can be produced without inflicting significant damage on the environment.

This is the approach taken, for example, by researchers at the Stockholm Resilience Centre, who define a set of nine 'planetary boundaries', within which humanity can operate safely, but transgression of which could trigger harmful environmental change.[337] Out of these nine, we have, they calculate, already crossed three: climate change, biodiversity loss and nitrogen depletion. This way of thinking about natural limits is clearly much more nuanced than Malthus', but it does mean that a calculation of how much food we could produce, and whether our current production is nearing the natural limit of the planet's productive capacity, is no longer straightforward or obvious. Indeed, since it immediately has to take into account not only space but greenhouse gas emissions, soil nutrient depletion, water resources and the effects of deforestation, it becomes extremely complex. These factors, while immeasurably complicating the calculation of the limits of possible planetary food production, are at least related to the extent of our physical resources, but they aren't the only ones. If we imagine trying to calculate the total possible food productive capacity of the planet, it immediately becomes apparent that socially-determined boundaries are just as important as the physical expressions of intolerable environmental strain.

Socially-determined environmental boundaries: fitting capitalism on the planet

We have already seen in the previous chapter how signs of problems in food production like famines cannot be taken as signs of an absolute physical disparity between the amount of food produced and the numbers of people needing to be fed. While this might seem to be a common-sense reason for famines to occur, in fact the history of famines shows that they are rarely if ever about an absolute dearth of food, therefore affecting everyone in a society. While there is usually some food production problem (potato blight, harvest failure etc.), famines are socially-constructed events, in which who lives and who dies is determined by social rather than physical factors. If, for example, Ireland had had a different social system in 1845, while the potato blight may still have arrived there, the effect would have been very different.

When thinking about food production, a famine is the most obvious imaginable sign that a natural limit has been reached. An understanding that famines have to be viewed within the context of the social structure within which they occur also indicates that natural limits themselves can't be abstracted out of the societies operating within them. Planetary boundaries act on societies, not simply numbers of people attempting to live on the planet, or indeed in a particular area. An example of a society considered to have crossed its natural boundaries – exceeded its carrying capacity – can illustrate the point.

The Maya are often cited as an example of a society destroyed by ecological collapse. The Mayan civilisation, a world of competing cities occupying the Yucatan peninsula in modern Mexico, thrived from around 200 AD until the ninth century, when one by one the cities appear to have been abandoned. The suggestion is that this collapse was the result of climate change, specifically a series of devastating droughts in the ninth and early tenth centuries. The Maya were used to dealing with water shortage, living as they did in a dry region, but the period in which they flourished had been fairly wet in comparison to the drier conditions beginning in the ninth century. More pertinently for considerations of what lessons we could learn from the Maya, it's also argued that they made themselves vulnerable to drought because over the six hundred years of the classic Mayan period, the population had expanded to the extent that it was 'operating at the limits of the environment's carrying capacity'.[338] The evidence for the droughts is compelling, and the archaeological evidence clearly shows the cities declining and then being abandoned. However, this doesn't necessarily mean that the problem was simply too many people for the available resources.

Richardson Gill, the archaeologist who coined the idea of the Maya falling victim to overpopulation and drought, argued that millions of Maya would have died of starvation and thirst, but this isn't necessarily the case. While the population of the area in the eighth century was probably at an all-time high, there were still millions of Maya people living there after the cities had been abandoned. The collapse of the Maya regimes could have been just that, the end of a particular society, rather than a full-scale destruction of a people. It's also noteworthy that smaller, less powerful cities in the drier north of the Mayan territory seem to have managed to survive the drought, significantly at the same time as they seem to have experienced a shift in power away from the royal authorities to the merchant class.

The point is not that the drought would not have posed immense difficulties for large numbers of people living in the Maya territories, nor that the collapse of the Mayan cities did not entail some reduction of population. It is that it was the society in which the Maya lived which could not be accommodated within the natural limits set by the drier climate, not the numbers of people *per se*. The Mayan cities in the south were vulnerable in a way that their northern neighbours do not seem to have been, not because they were living closer to natural limits but because their rulers, intent on pursuing dreams of conquering the other cities, were spending their resources on war rather than on maintaining the complex water systems on which the cities had relied for centuries.[339] Mayan society does indeed appear to have been destroyed by ecological collapse, but in the sense of ecological factors acting on a particular society, constructed in a particular way. A different type of society might well have survived it.

That the possible extent of modern food production is socially determined just as much as the Mayan response to drought becomes clear as soon as we try to quantify it, even in the simplest, Malthusian terms. The answer to how much food the planet could produce is of course dependent at a basic level on how much land could be used for food production, which raises questions not only about water resources, greenhouse gas emissions and so on but also about land ownership. We can only attempt to define planetary carrying capacity once we decide whether that is carrying capacity constrained solely by physical sustainability issues or by current land use and property rights. If we are calculating on the basis of existing property rights, we would have to exclude land used for non-food production, like golf courses, for example, and factor in their water use in our estimation of the water resources available for agriculture. This is usually done in estimates of available agricultural land, so a recent calculation in *Nature* has it that 38% of the Earth's land surface is used for agriculture and much of the rest cannot be brought into cultivation because it has other uses, ranging from urban land to nature reserves.[340] The point is not that these other uses are necessarily illegitimate, but that they are products of the social system. Under a different social structure, the golf course which is sacrosanct now could go from private property unavailable for crop-growing to prime agricultural land. The amount of land and its physical potential for cultivation would not have changed, but the social structure in which it might or might not be put to use would determine its productive capacity.

This is not a solely theoretical question. The idea that land in parts of east Africa has been destroyed by overgrazing, for example, is a commonplace one, and east Africa often appears in lists of those parts of the world in which people are already being badly hit by the effects of climate change.[341] There is no argument about the catastrophic effects of drought in the region, but arguments about overgrazing in the area need to be viewed with the understanding that this is an argument about land use. The Masai have been being accused of overgrazing and destroying the land on which they have lived since the nineteenth century, when the British imperial administrators were open in their belief that the two options for the Masai were that they should 'either alter [their] habits or disappear'.[342] The problem was not that their way of life was actually damaging their environment, but that it appeared unproductive to the colonial authorities, and failed to fit into the Western duality in which land is either brought into intensive use or fenced off as a nature reserve. Viewed in one way, the Masai lands in Kenya are an example of how meat-eating is unsustainable for the planet. Without the colonial appropriations of grazing land for nature or hunting reserves, and without the effects of British imperial rule and globalisation, the Masai way of life could be seen as a sustainable strategy. The problem with Masai herding was not that it was destroying the land, but that it did not fit with the capitalist British Empire.

In considering the carrying capacity of the planet, what is under consideration is therefore not the number of people who could physically be fed but the ability of the planet's resources to support modern capitalism. It has been pointed out that the carrying capacity concept itself does not take account of ways of life like pastoralism which are only accommodated within capitalism with difficulty, as is clearly the case with the example of the Masai.[343] If capitalism were a particularly efficient way of supporting the world's population in reasonable comfort, this would be a distinction without a difference; it is because this is not the case that the fact that physical boundaries act on social systems, not numbers of individuals, is so important. It is the gap between capitalism's use of natural resources and a baseline use to support the global population which suggests that a different system could maintain that population without the environmental destructiveness which capitalism brings with it.

Food waste under capitalism

That modern Western food systems entail a considerable amount of waste is not a new idea. It has been estimated that in the US, 50% of all food produced is wasted,[344] while in the UK, the Waste and Resources Action Programme (WRAP) concluded in 2009 that the average UK household wastes 22% of all food and drink purchases, 81% of which could have been avoided. (The definition of waste includes all food and drink discards, including inedible elements like egg shells or meat bones, so the waste is categorised as 'avoidable', 'possibly avoidable' or 'unavoidable' to distinguish between items which could have been consumed and those which would always have been thrown away.)[345] At a global level, it's possible that as much of a third of food produced for people to eat is not used.[346]

This level of waste would clearly have a considerable carbon footprint of its own; according to WRAP, the UK's food waste accounts for 20m tonnes of greenhouse gas emissions a year, 2.4% of the UK's total.[347] On this basis, waste seems to be a major contributor to climate emissions, although it has even been claimed that an extensive programme to address waste, including using the land currently used to grow surplus food for reforestation, could potentially offset up to 100% of current greenhouse gas emissions.[348]

The waste issue also seems a particularly contemporary one. Even using the more conservative understanding of food waste's contribution to greenhouse gas emissions, it apparently represents an easy opportunity to make a difference: teaching consumers to use up their food by the sell-by date seems much less of an uphill battle than persuading them to eschew meat or motoring. With the economic crisis from 2008, wartime messages of waste elimination also began to seem not only newly relevant, but an essential part of the response to economic and ecological devastation. *The New Home Front* by Caroline Lucas MP and Andrew Simms, for example, set out staples of Second World War exhortations against wasting food or fuel, or the 'SquanderBug' which 'causes that fatal itch to buy for buying's sake – the symptom of shopper's disease' as reflections of how these principles could inform how we approach our current crises.[349]

The measures proposed in this particular report are a mixture of individual and collective actions, ranging from campaigns to enable people to learn to mend and reuse rather than throw away, to government programmes to bring empty buildings back into use. However, while this is indeed a sensible approach to reduce waste within the current system, as an assessment of the wastefulness or otherwise of capitalism, it isn't particularly helpful. In trying to work out whether it is our numbers or our mode of production which doesn't fit on the planet, the cause of the well-documented wastefulness in the food system has to be identified. It does, after all, make a difference to the chances a different system would have of avoiding exceeding natural limits if the waste arose from factors inherent in the system or from laziness and greed inherent in human beings.

Fighting the supermarkets

The most visible elements of the food system are of course

consumers of food and the places in which they buy it, and it's not surprising therefore that discussions of waste have focused on individuals and on supermarkets. The questions of which of these two is considered the main source of waste, and how that waste arises, are, however, important.

That supermarkets have a malign effect on the food system and on local communities is frequently recognised, with discussions of their role in, for example, killing off town centres in favour of out-of-town retail parks, driving small food producers out of business or eliminating varieties of fruit and vegetables through an insistence on a particular size, shape or colour. In discussions of food miles and their contribution to climate change, supermarkets also appear, not unjustifiably, as the bad guys. Not only are supermarkets seen to be driving the development of air-freighted, out-of-season produce, but the development of 'just-in-time' delivery by the big supermarkets in the last decade has clearly made a significant contribution to road transport mileage. In 2004, up to 40% of all lorries on UK roads were involved in food distribution, following a couple of years in which Asda, M&S, Tesco and Nisa had all increased their lorry fleets by about 20%.[350] This model of delivery, reliant as it is on frequent lorry trips to and from the supermarkets' delivery hubs, has also contributed to a shift away from rail towards road freight, as unpredictable, varying loads, decided at short notice, are difficult to deliver using rail freight systems in which space has to be booked in advance.

Supermarket chains have enormous power over their supply chains, providing such significant markets for food products that they are able to impose more or less whatever conditions they choose on suppliers. It is this control over the supply process which leads to homogeneity in fruit and vegetables, as supermarkets insist on conformity to certain sizes and appearance, and also to a significant amount of wasted food. Malformed potatoes or unsightly oranges, rejected by the supermarkets, may be perfectly fit to eat, but the growers may not be able to find another market for the supermarkets' leavings, and they may well be thrown away. In a similar example, suppliers of ready-made sandwiches to supermarkets deal with extremely perishable ingredients, but are faced with supermarkets who refuse to confirm their orders until less than a day before delivery, so that they can respond to last-minute contingencies like changes in the weather which could affect demand. Since the suppliers are forced effectively to guess how much of their product they will be able to sell, there is a clearly a considerable amount of waste built in here, especially as some supermarkets make it a point of their contract that the supplier is not allowed to sell food the supermarket rejects to other markets.[351]

In their advertisements, the big supermarkets portray themselves both as wonderful places to shop and as hubs of the community. A recent, infamous example was the advertisement for one chain which was shot in a market where in reality the last fruit and vegetable stallholder had just been driven out by the supermarket.[352] In the real world, supermarkets are the targets of a variety of different forms of community activism. On the waste front specifically, they are challenged along with sandwich chains by activists for the amount of food they dispose of at the end of the day, and the difficulty in getting any of this diverted to useful causes, like feeding homeless people, even though much of it would still be fit to eat. Ongoing 'freegan' activity to use the food the supermarkets waste, versus the supermarkets who may well instruct their staff to pour paint over discarded food, and who may even prosecute people found taking food out of their bins,[353] helps to keep supermarkets in the forefront as a significant problem when it comes to food waste. For many freegans, this is the point, taking food from supermarket rubbish bins being 'the propaganda of the deed', inevitable in the sense that in a consumer society, struggle would always emerge over consumption practices.[354] Their very existence is also increasingly disputed. The news that a supermarket chain is planning to open a new shop in their area is a prompt for many communities, not to break out the bunting, but to organise and resist. One of the best-known of these fights was the battle of Stokes Croft: the campaign against Tesco in inner-city Bristol.

I talked to Chris Chalkley, the Chairman of the People's Republic of Stokes Croft (PRSC), in a shipping container in the outside workspace at the Jamaica Street studios where the collective is based.[355] It's a setting which sums up much about what makes Stokes Croft unique, from the Bristolian nautical connection to the general sense that you are in a place which has always, as Chris says, found an alternative way of doing things. Situated just outside Bristol's medieval city walls, the area had long been a place for trading outside the taxable reach of the city fathers, but after it was heavily bombed in the Second World War, it became a site of urban dereliction, known for high rates of poverty, deprivation and drug use. On the other hand, it also had a large creative and squat community. The creativity in the area remains apparent. Not knowing Bristol well, I was worried on the bus from the station that I wouldn't know when I had arrived, but when it came to it, it was obvious. I didn't need the automatic announcement of the next stop, all I needed to do was get off when I saw the buildings covered with street art and the passers-by with green hair or dreadlocks.

Chris himself comes from the south-west, but hasn't always stayed there, travelling as far afield as the Australian outback, where he worked on wheat and sheep farms for a period in the 1970s. He came to Stokes Croft in the mid-1990s, when the crash in property prices enabled him and a group of friends to buy the building which is now the Jamaica Street studios. The People's Republic of Stokes Croft, he tells me, was inspired by his experience in 2006 when he decided to paint on the fence at the property. People kept coming up to him while he was painting, asking if he was allowed to do it. In this case he was, as it was his fence, but he realised that their questions came from their experience as local street artists whose murals were being diligently painted over with grey squares by council workmen as fast as they could create them. The council was only a mile and a half distant, but they were too far away to understand the needs of this creative area and were only interested in damping it down. What was needed was a way of building change from the bottom up, a way to bring together people in the area who wanted to question the *status quo*, and so the People's Republic of Stokes Croft was born.[356]

Since then, the creative side of Stokes Croft has become well-known, and there is a sense that it is going from an embarrassing spot of urban decay to part of an advertisement for vibrant Bristol. The council are trying to get in on the act, paying around £250,000 to put street art on a number of council and corporate buildings in the area, and I noted in late 2013 that among the buildings sporting street art along the main road was a mainstream-looking solicitors' office. Chain stores don't tend to be keen to locate in very rundown areas, and the rough reputation of Stokes Croft had kept the multinationals out. This was indeed part of its charm: when the area was designated a conservation area in the 1980s, it was because it was characterised by a mix of properties and small businesses. However, as Stokes Croft became less 'rough' and more 'interesting', this came under threat.

In November 2009, Tesco filed for planning permission to open a Metro store on Cheltenham Road, on the edge of Stokes Croft. ('Some people', Chris said, 'would call it Montpelier [the neighbouring part of Bristol].' He paused. 'With good reason.') The application was lodged in the name of a Bath solicitor, so the 50 or so local residents who received the notices weren't concerned and the council received no objections. In February 2010, a resident got chatting to the workmen who were starting to refit the empty property and discovered to their horror who the real clients were. The fight to stop Tesco became a rallying point for the entire community. The campaign delivered 2,500 signed flyers opposing the Tesco to the council and went door to door to poll residents on their view, which showed around 90% were opposed. After the building was squatted to stop Tesco from moving in, there was, in Chris' words, 'a big old ruck' where the police found themselves standing between the local population, who didn't want the multinational, and the 'corporate thugs'. The site was like a military camp, surrounded by steel fencing and occupied round the clock by security guards, while the workmen finished and the anti-Tesco campaign fought a rearguard action on everything from shopfront signage to refrigerator size.[357]

In the end, however, the council voted by four to three to allow the shop to open. Refusing permission would have been an act of courage which would have set an example for councils everywhere, but one which the council just weren't prepared to perform. As one (unnamed) councillor said to Chris, outside the barricaded Tesco, 'We dare not refuse Tesco planning permission, because if we do, they will appeal, and they will win on appeal, and then I will have wasted massive sums of taxpayers' money'. The shop opened in 2011, only to be trashed in the riots which broke out in Stokes Croft, as in much of the UK, in August of that year. The riot in Stokes Croft had complex causes, but it is known locally as the 'Tesco riot', in recognition of the issue which had brought the community together and out onto the streets.

The active phase of the campaign against Tesco in Stokes Croft may have ended, but the work of the PRSC continues. What is important, Chris says, is that people have to take on the issues in their local area, and start thinking about not only what they dislike about the current system, but about what they do would do differently. Bristol is increasingly at the forefront of attempts to build an alternative food system and possibilities for the future include a food hub for Stokes Croft. In the meantime, on the wall overlooking the Tesco, a huge mural reminds would-be shoppers to 'Think local, boycott Tesco'. It is not a monument to a failed campaign, but a continuing engagement with the community which makes sure that the presence of Tesco does not become normal. Chris tells me that he doesn't shop in any supermarkets any more.

The battle of Stokes Croft was one of the more prominent campaigns against a supermarket moving into a community, but it is far from the only one. Indeed, when I visited Stokes Croft, I received a sense both of the radicalising power of campaigns like the fight against Tesco and of how widespread these can be. Chris' mother, a sprightly lady in her 80s, dressed from hat to shoes in scarlet, was at the PRSC helping out in the shop. I explained why I had come and she told me enthusiastically about the campaign she was involved with where she lives in genteel Cheddar, in Somerset, against a Tesco opening. They had succeeded for the moment, she told me, although Tesco still owned the land so they hadn't relaxed yet. The only unfortunate downside to the Cheddar campaign was that they had not managed to keep all supermarkets out; while the campaign against Tesco was at its height, Sainsbury's had managed to sneak in, almost unnoticed, on the other side of town.

When feeling against the effects of the supermarket chains is so strong, it would seem obvious that popular discussions of the problem of food waste, and its possible solutions, would focus around the wasteful practices of Tesco, Sainsbury's *et al*. It is interesting therefore to note how supermarkets do not, in fact, appear as the primary villains in many discussions of food waste. In part, the focus on individual consumers and their buying and eating decisions mirrors the shift in discussions of food and climate change more generally away from issues of transport and packaging towards seeing the problem as inherent to certain types of food. It is notable, for example, that while plastic packaging would have appeared in the past as one of the major environmental sins committed by the supermarkets, with high-profile campaigns like the Women's Institute's packaging day of action in 2006 encouraging consumers to reject excess packaging,[358] now, according to WRAP, packaging waste is comparatively less significant than the waste of the food itself. In other ways also, discussions of food waste reflect the overconsumption view of food and climate change.

The wasteful individual as bad guy

Households are only responsible for about a third of all food waste,[359] but discussions of the waste issue have a tendency to become discussions of individual wasteful behaviour. So, for example, Gordon Brown made headlines in July 2008 when he accompanied the launch of a Cabinet Office review of food policy with exhortations for people to throw away less food,[360] while the role of supermarkets in food waste was discussed primarily in terms of a debate over whether banning 'buy one get one free' promotions, otherwise known as 'encouraging unnecessary food purchases', would be a good idea.[361] In this view, food waste arises because we all buy too much food: 'In truth, we have simply become lazy and negligent about food, and we are blind to the true costs of wasting it'.[362]

This comment points to a common assumption about the source of food waste: that food is wasted because it is cheap.[363] It is difficult to find empirical proof for this assumption, reflecting as it does a market view of the world in which everything is valued by its end-users according to what it has cost, and which also appears in other contexts, like debates about so-called 'disposable' fashion from cheap outlets, so that increases in clothing waste are dubbed 'the Primark effect'.[364] Indeed, as well as arising from cheap food, with a neat circularity, food waste is sometimes also presented as evidence for the existence of an oversupply of cheap food. You would think that an oversupply of food was by definition waste, but what is meant here seems to be not that food is produced and wasted through not being sold, but that it is so cheap that consumers buy it almost so that they can throw it away in an orgy of consumer indulgence.[365]

The concomitant of the idea that cheap food enables us to buy things we aren't going to consume is, of course, that we might consume them even though we don't need them. 'Overeating to obesity' is, as discussed in chapter three, a central part of current definitions of overconsumption, in which sustainable eating is eating only the daily recommended calorific requirement and not a mouthful more. The progress of the wasteful individual, from supermarket to dustbin, is neatly summed up by Ian Roberts and Phil Edwards in their *The Energy Glut*: supermarket free-parking encourages us to drive there, and because we don't have to carry it home, 'consequently, we will buy a lot more food than we need. We will eat some of it and waste some of it. The extra food we eat will make us fat and the food that we waste will be dumped in landfill sites where the rotting vegetable matter will release methane, a potent greenhouse gas.'[366]

This model allows considerable persuasive power to food retailers, whether it is their buy in bulk offers or their cunning car parking, but it is still the responsibility of the consumer to resist them in order to avoid waste. The supermarkets are, it seems, encouraging their customers to give in to their most base urges, in Tim Lang's phrase, 'colluding' with them,[367] in what begins to seem a struggle between the id and the superego, played out in the aisles. The notion that we will waste as much food as we can unless strictly regulated seems to borrow from the dieting canon the assumption that uncontrolled behaviour when it comes to food will always be excessive and destructive. It is important because it locates the origin of food waste squarely in the innate properties of human beings, not human beings under capitalism.

Wastefulness: biology or capitalism?

Roberts and Edwards make this particularly clear: according to them, we are programmed to be this way by our evolution: 'The desire to eat whatever food is available and the ability to store excess energy as fat is hard-wired into our biology'.[368] This is not the place to discuss how dubious is the notion that any behaviour is 'hard-wired' into our brains, nor indeed the unfortunately pervasive idea that modern social norms can be traced to *homo sapiens'* history on the African savannah. It is sufficient to note that this interpretation places the consumer and their desires at the heart of the food system, while the supermarkets, however cunning and tempting, are of secondary importance in terms of waste.

This seems a considerable underestimation of the importance of supermarkets in food waste. They aren't simply places where we can overbuy, but stores whose business model is based on procedures which entail wastage. The constant availability of a wide range of products, whatever the time or season, is a significant way in which the supermarket chains compete with each other: when Julie Hill, author of *The Secret Life of Stuff*, suggested to a supermarket senior executive that 'the best thing the store could do for the environment was to run out of food from time to time... this was met with a frosty silence'.[369] The obvious rejoinder is that supermarkets are pushed into this by the demands of their customers, by 'the abundance model [which] has been built into consumer culture',[370] but this is to concede too much to the idea that demand drives supply. It is true that Tim Lang, in this quotation, is viewing this consumer culture as the creation of supermarkets among others; the passive building of the abundance model was carried out by the businesses now benefiting from it. But the notion that even if started by business interests, a consumer culture of demand for constant gratification compels businesses to satisfy it, overestimates consumer power compared to business power in this scenario. That consumers ever clamour to be able to buy strawberries at 3am in January is debatable, and it is difficult to sustain a convincing argument that it is such desires on the part of consumers which lie behind the food waste problem.

It is easy to see why figures like Gordon Brown might be in favour of the view in which the market is merely responding to what consumers want, and could be brought to act differently if consumers managed to have different, less wasteful demands. Other commentators, less wedded to the free market than our ex-Prime Minister, have varying degrees of faith in the power of consumers to change supermarket practices, ranging from the comment that 'Supermarkets do actually listen to their clientele',[371] to suggestions for government regulation so that rather than giving consumers the choice of a green product, every product would be a green choice,[372] to the disciplined circumvention of supermarket blandishments recommended by Roberts and Edwards. While there are various prescriptions, what they share is an underlying view of the origins of waste which holds that individuals are the source of waste as a result of their wasteful behaviour; and that however much this is encouraged to serve the needs of consumer capitalism, it would presumably be a problem to be strenuously overcome regardless of the system they were in. The implication of the concentration on the wasteful individual is that, whether waste could be dealt with without changing the system, or whether the conclusion is that the system needs to change, human greed and laziness would always have the potential to create similar problems.

If this were the case, we would have to conclude that, while we might be able to trace waste at points in the food system to the particular workings of the market under capitalism, the issue for natural limits would be not capitalism, but human behaviour. If we have a culture of wasteful abundance which puts strain on our finite resources because humans are programmed to it by their evolution, it would be difficult to argue that any system with which we might replace capitalism would be significantly less wasteful than the system we have now. We would be back, in fact, at a Malthusian conclusion in which the existence of humans who give rein to their desires is the root of the problems for the planet. The nature of the profligacy may have shifted from breeding to eating, but in borrowing the dieting world's fear that people (especially fat or formerly fat people) will eat the world unless prevented, discussions of food waste which locate the problem with the individual show themselves to be surprisingly close to the views of Malthus when it comes to carrying capacity. In order for them not to destroy the world, working-class people have to recognise their profligate urges and exercise prudential restraint. Now, where have we heard that before?

If we conclude that we see waste within capitalism as arising from individual behaviour, rather than from aspects of the capitalist system itself, it should follow that other systems would also suffer from problems of waste and overconsumption. As we have seen from the example of the Maya, there have been environmental crises throughout human history, and it is possible to argue that various human societies have had significant deleterious effects on their environments. The role of various pre-modern societies in significant deforestation, from Easter Island to Celtic Britain, is just one well-known example of how different cultures have arguably damaged their environments, long before the development of capitalism. There is an argument, in fact, that we should consider the environmental destructiveness of capitalism not because it is a uniquely destructive system, but simply because it is the system we have. Our concentration on capitalism, in this view, should not blind us to the possibility that other systems could be just as destructive.[373] The wastefulness of the capitalist food system would therefore not necessarily be solved by replacing capitalism, but would represent a way in which we would be up against our natural limits whatever system we were in.

Industrialism as the source of environmental strain

It is clear, of course, that industrial societies have placed much more strain on their environments than non-industrial ones, and climate change is easy to understand as a crisis of industrialisation. In food production specifically, it is again a specifically industrial food production system which can be seen to have brought us to the limit of the planet's productive capacity. Broadly speaking, it is industrialisation which has enabled food production on a scale which has created the modern 'cheap' food phenomenon. That average household spending on food in the West is much lower than in the early twentieth century is an often-repeated fact, (although it is also one which should be treated with some caution, since it fails to take into account the possibility that other calls on the household budget might also have changed in price, for example housing). If consumers waste food because it is so cheap, this line of reasoning could force us to conclude that it is industrial society, rather than capitalist society specifically, which does not fit on the planet.

The Industrial Revolution arose, of course, within capitalism. It owed its existence to the way in which the transition from feudalism to capitalism had allowed the accumulation of capital to then be invested in the development of industrial machinery. The fact that industrialisation was a creation of capitalism does not mean that industrial civilisation could not exist under any other system, but whether industrial production would always inevitably be as destructive as under capitalism is a deeper question.

It is also a rather unfashionable one. It was a principle of the Stalin-era and later Soviet Union that it was capitalist relations of production which caused environmental harm, and that this could not possibly apply to their industrial development. This was clearly not the case, and the legacy of pollution and industrial contamination over large areas within the former Soviet Union, highlighted by but not restricted to the Chernobyl nuclear accident, has shown how damaging the effects of Soviet industrialisation were. There is no shortage of comments on how wrong it was to assume that 'socialist' industrial production was immune from the environmental damage inherent in capitalist production, so for example, this statement from a Soviet commentator in 1980 was quoted by a Rand Corporation writer in 1993 as a demonstration of how tragically blinded by ideology Soviet thinkers had been: 'Like other global problems, those of ecology have a social origin, their solution largely depending on the character of the social system... The socialist states and the communist parties proceed from the conviction that the socialist system offers the optimal possibilities for resolving these problems.'[374] However, the question is not *whether* the Soviet authorities were wrong about the potential for environmental harm from their industrial production – as they patently were – but *why* they were wrong.

It could be concluded from the Soviet example that it is industrialisation and not capitalism which is particularly destructive to the environment. An important development of industrialisation is the transfer of production from human hands to machines, which enables enormous increases in the amount of possible production. Resource extraction by people with hand tools is limited to their labour power, maintained by the resources available to feed them, but once machines are introduced, the increases in labour productivity mean that many more resources can be extracted for use: production is no longer limited to the unaided labour power of those working on it. It is this which has enabled capitalism to use energy stored from past eras, for example by mining, as well as using the immediate energy which we are getting now, ultimately from the sun.[375] If the scale of production which industrialisation enables is seen as the origin of our environmental problems, the solution would have to involve changes in that production. The implication of this line of argument is that a society existing within the planet's natural limits would have to be a post-industrial one, whether or not it was also post-capitalist. In terms of food production, this would have severe implications for how many people could be fed sustainably, given the scale of food production enabled by industrialisation.

However, it is worth noting that the Soviet regime cannot be taken as evidence that the problems of industrial production go beyond capitalism. Soviet Russia and associated states should more accurately be described as 'state capitalist', not as socialist; their industrial development is not a demonstration of how industrial societies might function outside capitalism.[376] While the historical development of industrialisation cannot be separated from capitalism, this does not mean that industrial production of any type would always be as transgressive of natural limits as current industrial production under capitalism. In fact, the particular workings of capitalism make it uniquely profligate of resources and destructive of the environment, for reasons which are not simply related to the technology of which it has enabled the development.

The problem with industrial production from an ecological point of view is the intensive resource use which it enables. This is an uncontroversial statement, but it carries within it the assumption that in human societies, resource use will always increase to the maximum permitted by the technology available. In other words, we will always exceed the limits of our environment if we can invent a way to do it. This seems to return us to the idea of inherent greed, or at least profligacy, of humans as the fundamental problem. On examination however, the assumption here is not about the fundamentals of human nature, but about capitalism.

The unique destructive power of capitalism

In discussions of overconsumption from Galbraith on, it's accepted that modern capitalism is associated with growth, but it tends to be presented as a lifestyle choice which capitalism could learn to do without. Growth is something which capitalism is 'obsessed with', or 'makes a fetish of'. It is a familiar form of argument in environmental issues – journalistic descriptions about how as a society we're 'addicted' to oil, for example – and the problem in all its manifestations is the same, that it reduces an aspect of the structure of a society to the individual proclivities of its members. Capitalism is not 'obsessed with' growth because particular capitalists believe that continual growth is essential, any more than we are dependent on oil as a result of individuals' continued car ownership. Growth is part of the structure of capitalism; indeed, without continued expansion, capitalism does not work at all.

Capitalism is not a static but a dynamic system, which works only on the basis of ever-increasing profits. Competition between capitalist businesses impels each to seek expansion to outdo their competitors – the drive to greater and greater accumulation – but the profits generated by this are effectively dead capital if they cannot themselves be invested in a way that will get the capitalist a compound return on their investment. It is this underlying dynamic in capitalism which drives the cycles of boom and bust, expansion and depression, of which we've all had a potent reminder in the last few years. These aren't a malfunction of capitalism but simply a result of how the system works. Because the need for growth arises from the need for competitive advantage, an explicit belief or not in the value of growth within the system will not change it: even chief executives who talk the talk on green issues are still looking for ways to become more profitable than the other businesses in their industries, and to beat them to new areas into which to expand. Industrialisation comes in here because mechanised production initially allows those capitalists who adopt it to produce larger quantities of products more cheaply than their competitors, although it also contains the seeds of the next cyclical 'bust' by causing the rate of profit to decline. Marx showed how this happens as companies invest their profits – the surplus value produced by the labour of their workforces – in machinery and thus convert it from productive capital. As the rate of profit falls, this is another driver of expansion, but is solved until the next cycle when many businesses in a sector collapse, and their surviving rivals improve their rates of profit by eating the corpses.

The adoption and development of mechanised production under capitalism is then undertaken in search of competitive advantage: any benefits to society in general, in terms of cheaper food production, improved health technology and so on, are as incidental as the damage caused by industrial production is unconsidered. A system in which industrial production was employed for the good of society, rather than for the profit motive, might have a different environmental effect, particularly if the need for profit was not also standing in the way of the development of technologies to avoid greenhouse gas emissions. The problem is not, however, purely one of scale: while a lesser volume of capitalist production would be less harmful, it is clear that the damage which capitalism causes to the environment is not in linear relationship to its size.

The problem with capitalist production is not merely quantitative, but qualitative. As John Bellamy Foster recently pointed out, 'it is easier for the system to grow by producing depleted uranium shells to be used in imperialist wars or by expanding agribusiness devoted to producing luxury crops to be consumed by the relatively well-to-do in the rich countries than it is to protect the integrity of the environment or to provide food for those actually in need'.[377] In this example, the division is between undesirable activities which involve the production of profitable goods, and the more socially-useful, which do not, but this is not the only qualitative problem with capitalism.

As Marx identified, the birth of capitalist agriculture in the nineteenth century entailed the transformation of aspects of the natural world into commodities, whose value was not their use-value but the profit they represented for the capitalist. As industrial agricultural production grew, driven by the search for greater profits, and as the separation between the industrial town and the countryside widened, agriculture in western Europe ceased to be a self-sustaining system. A metabolic rift had developed, as the scale and techniques of industrial agriculture removed nutrients from the soil without replacing them. Because the industrial system did not try to find ways to reuse and replace these nutrients, they were instead allowed to become pollutants as urban waste, while in the meantime the soils were becoming exhausted. The solution, in capitalism, was not to review the way in which industrial agriculture was developing, but to replace the soil nutrients which were progressively being lost by extracting them from elsewhere. The nineteenth century saw the development of the Peruvian guano trade, in which vast amounts of guano deposits were imported from Peru to Europe to serve as fertiliser. This removed natural resources from one part of the world for them to become a pollution problem elsewhere, at enormous human cost to the workers on the guano and those who suffered from the associated destruction of the Peruvian economy through the involvement of Western governments in the country.

The history of the Peruvian guano trade is an important illustration of the consequences of the metabolic rift. It also illustrates how it is that capitalism has such destructive power compared to other modes of production. It's easy to assume that the global reach of capitalist business in the last two centuries is simply a product of improved transport and communications, but in fact this is a key attribute of capitalism. Western imperialism has always been driven by the needs of capitalism for expansion, for new sources of cheap labour as a result of the declining rate of profit, or for new natural resources to exploit. As the history of imperialist intervention around the world has demonstrated, this exploitation of labour and natural resources has not tended to include concern for the long term effects on the society and land: capitalism's model is to take what it wants and move on. It is this which is at the heart of capitalism's unique ability for large-scale destruction. Elites in other modes of production may not care any more than capitalist elites do for the long term wellbeing of the people they dominate, but they have perforce to take more care of the natural resources they hold. The difference is because of the mobility of capitalism.

Commodification means that very different products are essentially exchangeable: a capitalist can make profits from, for example, palm oil production in Malaysia, but can take those products and invest them in in a completely different sector on the other side of the world. Under any other mode of production, members of an elite who were enriching themselves from palm oil production would have to care about the prospects for continuing that enrichment in the long term, or their days as a wealthy elite would be numbered. It is only in capitalism that those same elites can extract their profits and move on, leaving others to worry about the mess left behind them. This is not to argue, of course, that rulers in other modes of production have been universally ecologically-minded, if only from self-interest. Far from it: many human systems other than capitalism have managed to destroy or at least threaten the environments which sustained them. Soil exhaustion, for example, was not a problem first known in western Europe under capitalism in the nineteenth century, but had been an issue at various times under feudalism as well. The point, however, is that elites whose power ultimately rests on their possession of particular land, and exploitation of the labour of particular populations, find themselves in trouble if those lands cease to be productive. In capitalism, global elites have learnt that their resources can also be global: their power is not based on the control of particular areas or resources, but of capital itself. There is no inbuilt need for them to take care of any particular environment, and no obvious short-term penalty if they fail to do so. That the consequences only arrive when the entire global environment is under threat is the root of capitalism's destructive power.

Food waste as a source of profit

The conclusion that it is all capitalism's fault is not a particularly rare one in left environmental circles, but it is important to take it beyond this generality. If capitalism in general terms is a system with unique potential to cause resource depletion and environmental damage, how the global food system has developed under capitalism is key to the ecological problems it is now causing. To return to the issue of waste, for example, it is evident that food waste is not just an unfortunate side effect, but an essential part of the supermarket food system. This might seem counterintuitive, and can indeed be portrayed so, since wasted food means on the face of it a wasted opportunity for profit. Indeed, one suggested approach to dealing with food waste is to encourage supermarkets and companies to deal with it to improve their bottom line, with the environmental benefits of waste reduction as a fortunate add-on.[378] It is difficult to imagine, however, that supermarkets, suppliers and producers are unaware that food waste costs them money until this is pointed out to them by environmental activists.

The problem is that eliminating waste while preserving profits is not an easy proposition, and it is not the case that the players in the food supply chain have been indifferent to waste up until now. On the contrary, supermarkets can be seen to be extremely concerned to reduce their waste and therefore their costs, but they are doing so not by giving unsold food away at the end of the day, but by using their power to insist to their suppliers that they will not confirm the amount of product they want until just before delivery. Suppliers with sufficient power will then try to pass this on by implementing similar 'just in time' policies with their own suppliers, and so on until the least powerful parts of the supply chain get stuck with food that can't be used.

What this highlights is that it isn't waste *per se* which costs the individual companies which are part of the food supply chain, but waste which they can't pass on. In their position as the most powerful companies, the supermarkets can benefit from the short lead times and speedy deliveries which would be lengthened if waste were cut across the entire chain, but they reduce the waste for which they have to pay by forcing their suppliers to bear the cost of it themselves. It would be difficult to argue that reducing waste in this scenario would increase the profits of supermarkets: within a market system, their behaviour is perfectly sensible and the levers to persuade them to change their practices are unclear. Similarly, it is not surprising that supermarkets are resistant to allowing their surplus food to be distributed for free: they run the risk of losing either customers who could pay for it, but chose to wait until it is deemed 'wasted', or the exclusive image of high-end products if they are given out at the end of their sell-by dates to those who wouldn't otherwise be able to pay for the privilege.

The waste caused by the supermarket system within capitalism is both serious and revealing about the workings of the system as a whole, but it is clearly not the whole problem. As John Bellamy Foster recently pointed out, the suggestion that we need to 'ecomodernize our shopping habits' hardly covers the ecological crisis facing us.[379] Waste of energy and food is not simply the preserve of supermarkets, nor indeed of consumers themselves, but is built into the entire food system. This might seem an unexceptional statement, but the point, as with supermarket waste, is that this is not waste which happens because chief executives have not yet seen that reducing waste is in the interests of their bottom line, but waste which is such an integral part of production that it is responsible itself for a large measure of profitability.

Processed foods: an example

The development of processed foods is a useful example. As we have seen, for many, the degree of processing to which food has been subjected is now viewed as less important for its impact on the planet than its intrinsic nature. Processing food is often lumped together with the effects of transporting it as things which we used to be concerned about, but have now learnt are not as serious as meat-eating. However, to point out that meat production accounts for more greenhouse gas emissions than processing food is not the same as to say that the energy used in food-processing is negligible. In any case, the diet of the poor, fat people in the frame for climate change is usually characterised as containing large numbers of processed foods, and eating fewer of them generally appears as a step towards a more ecologically-conscious, as well as healthier, diet.

This view of processed foods casts eating them as a personal choice; a moral failing or a decision dictated by circumstances like lack of time or cooking facilities, depending on how condemnatory is the position of the viewer. What it precludes is an understanding that the spread of processed foods is neither the result of indolence, decline of cookery teaching in schools nor any other personal changes, but of the fact that our food system is capitalist. Processed foods are nothing more than the attempts of food companies to maintain and increase their profits in the face of the tendency of the rate of profit to fall. It is for this reason that they exist and that Western diets have become increasingly likely to include them.

The conventional narrative of the development of processed foods has it as a consumer-led revolution. As one account summarises it, after the end of the Second World War, 'manufacturers desperate to unload the newfangled frozen, canned and dried foods they had created for foxhole consumption found a willing population of women who had gone to work during the war, and who were no longer interested in fussing over meals'.[380] Ready meals, heat-and-serve foods and the like were bought enthusiastically by women who still had the main responsibility for feeding the family, but who didn't want to spend hours in the kitchen to do so, or did not have enough confidence in their cooking skills to cook from scratch. This story can be presented in a more or less critical way of the processed food-embracing consumer, depending on whether it's a nostalgic celebration of weird processed foods or an argument that we are eating ourselves and the planet to death, but the constant is that processed and junk foods were developed because that is what people wanted to eat. It's a simple tale, with only one unfortunate aspect: that it isn't true.

It's clear in fact that food manufacturers in the US after the Second World War had an uphill struggle to persuade women to adopt their products in place of their own home-cooking. It is true that the media in the 1950s proclaimed over and over again that convenience foods were the new foods of choice, but this reflected what food manufacturers wanted to be the case, rather than an actual shift in behaviour. The media stories, with their proclamations of 'a revolution in eating habits' or housewives 'waiting with outstretched arms' for processed food, were protesting too much.[381]

Betty Friedan's discussion in her 1963 *The Feminine Mystique* of how advertisers tried to sell food and cleaning products to women is interesting here. According to Friedan, advertisers of time-saving gadgets and products didn't see career women as their primary market, even though you would think that they would be the group who would be most in need of time-saving, as they were too sceptical and picky. The ideal market was the 'balanced homemaker', a woman who spent all day at home but who was trying to convince herself that 'managing a well run household' was a sufficient use of 'her own executive ability'. The way to sell processed foods to this group, the manufacturers worked out, was both to prey on their insecurities about the standard of their cooking, and to convince them that processed foods simply removed the drudgery from cooking, while leaving them room to express themselves through the finishing touches.[382] As Ernest Dichter of the Institute for Motivational Research put it, in an oft-quoted dialectic (Betty Friedan relied heavily on Dichter for her chapter on advertising, but, coyly, did not name him):

> Thesis: "I'm a housewife."
> Antithesis: "I hate drudgery."
> Synthesis: "I'm creative."[383]

That the food manufacturers were able to overcome consumer resistance is a demonstration of how little the shape of our diets is determined by consumer choice, unmediated by corporate interests, just as, for example, the shift in US diets from pork to beef in the same period was dictated by grain and beef producers.[384] The question, however, is why it was worth food companies' while to convince consumers to embrace processed foods. While canning and freezing were known before the mid-twentieth century, other processing techniques were largely developed to feed the troops in the Second World War, and military food has not usually been so appetising that it has been enthusiastically adopted by civilians. The answer, of course, is that processing equals profit.

The profit for the food manufacturer comes from the value generated by the labour on the food. A farmer pays labourers to harvest a field of potatoes, but he doesn't pay them the full value of the crop – he pays them as little as he can get away with, which if they're undocumented, migrant workers, is probably very little indeed – and the difference is the surplus value, the part of the value of the potatoes from which profits can be made. The farmer could sell the potatoes to a retailer to sell from a sack in front of the shop. The retailer would make some profit from the value of the shop workers in selling the potatoes, but as far as late capitalism is concerned, this isn't a particularly profitable chain. However, if the farmer were to sell the potatoes to a food manufacturer to wash, peel, dry and sell to another manufacturer to put into heat and serve shepherds pie, that would add another two stages for the generation of surplus value, and therefore profit. All of which explains why it's harder than it used to be to buy potatoes with the earth still on.

The more production steps food goes through, the more surplus value it represents, and so ultimately the more money the capitalist makes from it. So we have what John Bellamy Foster calls 'ever more baroque' commodity chains, in which food undergoes another processing step whose sole point is to generate another increment of profit, with the result that, in Foster's example, a muffin, which could be whipped up in a domestic kitchen in half an hour, goes through seventeen energy steps, accounting for twice as much energy in production as it takes to grow the raw material.[385] The expenditure on processing is utterly out of proportion to the value of the initial ingredients, but that does not matter. Capitalism, whatever we're told by its defenders, does not have to be rational. It only has to be profitable.

The energy used to process that muffin is not used because humans are naturally profligate and lazy, but because it is actively more profitable under capitalism for the muffin to go through seventeen energy steps than seven. Although the tendency in much recent writing about food and energy use has been to minimise the energy from processing compared to the energy inherent in the food itself, this is deceptive, as it underestimates the extent to which the need to add processing steps to add profits has shaped the entire food system. It is hardly surprising that a system in which energy use is rewarded by profits and simple production chains are effectively penalised appears to be structured around as much energy use as possible. It's because it is.

The consequence is that we can't conclude that we are transgressing natural limits as a result of the types of food we eat, or the number of people on the planet, when the profligacy of capitalist food production is its dominant characteristic. It's not so much that the current food production system is wasteful, it's that the wastefulness is the point. If we want to know if seven billion, or indeed nine billion, people can live on the planet without destroying it, or if we are up against a fundamental natural limit which no system could stay within, we need to consider if we could do so with a less wasteful system than capitalism. It is capitalism and not basic human profligacy which makes current food production so destructive, but could we really all have a decent standard of living without that destructive capitalist production? It is to a consideration of the alternatives, and of how we might get there, that we now turn.

6

Beyond capitalism, and how we might get there

One of the enduring myths of industrial capitalism is that economies of scale always apply. Production on a large scale will always need large-scale, industrial enterprises, otherwise the volumes required will be either impossible or prohibitively expensive. Thus, as far as food production is concerned, it can seem obvious that there is no way out of the current industrial system if we want to be able to feed everyone, let alone cover the projected increases in global population. Even the realisation that it is capitalism and not individual human greed which is the problem is not an automatic way out of the argument which holds that if everyone is to eat something, some of us have to eat less. This chapter could be a paean to the delights of savouring small quantities of food, the benefits of reducing obesity when everyone would have the calories they are deemed to need and no more, or of how our lives would be so much better without capitalism that limiting our food intake would seem only a minor inconvenience. It is not. The assumption that bigger and more industrial is always better is not a natural law but itself part of the capitalist system, and even within this system, can be shown not always to apply.

Bigger is not better

In agriculture, it's tempting to interpret larger fields and farms, and industrial operations devoted to production of a single crop, as more modern and efficient. This is part of the pervasive but ill-founded myth that labour-intensive, mixed farming is primitive and therefore wasteful.[386] It seems that 'modern' farming methods are not the way to get the best yields. In Brazil, for example, 30.5% of the agricultural land is occupied by small family farms, but they manage to produce 40% of the total national value of production.[387] It is true that very small farming units are also unlikely to be the most efficient, but the most productive size range for farms appears to be well below the industrial farming model. A comparison between Amish and Mennonite farms in the US with their industrial neighbours is instructive, as they are working with the same land and conditions, and the surprising result is that the non-industrial, small farms appear to be between two and ten times more productive.[388] The industrial tendency towards monocultures is also an unhelpful one, as the most productive farming maintains a mix of crops, and combines arable and animal farming, to make the best use of all the land available and maintain the best possible nutrient mix in the soil.

It turns out that turning farm output into just another selection of commodities, determining the balance of production in a corporate office according to which are most profitable, is not the most sustainable way to manage agricultural production. The metabolic rift which Marx identified in nineteenth-century agriculture as a result of urbanisation and industrialisation is perpetuated in modern industrial farming, so that for example the manure from farm animals, which could go to fertilise land for crop-growing, instead becomes an industrial pollutant. Fertilisers meanwhile still have to be bought in, with their own implications for greenhouse gas emissions and pollution, much like the Peruvian guano discussed by Marx. Recognising how far capitalist relations of production have been extended into agriculture points up how far it could be different under a different system. In the same way, understanding how large-scale industrial production suits the needs of capitalism, rather than the needs of the population, enables us to see the possibilities in small-scale urban farming. It is also the case that moving away from an industrial farming model for animal farming would enable measures to produce direct reductions in greenhouse gas emissions, since, for example, grass-fed cows produce less methane.

Sustainable farming in practice

The idea that farming in the West could be reclaimed from agribusiness and supermarkets is not just a theory, as there are numerous efforts at putting it into practice. Take Steve Green, for example. Steve decided to turn the yard in front of his house in Windsor, Ontario from 'a disaster of clay and grass' into an urban garden, and to blog about it.[389] It was a great success and in 2008 he went on to start the Windsor-Essex Community Supported Agriculture initiative: a farm on the outskirts of Windsor run by volunteer labour, aimed at supplying local people not only with access to fruit and vegetables but the chance to do something positive for their community. A grant from the Trillium Foundation followed and Steve now works full time on what has grown into a serious effort in urban farming.[390]

The Windsor-Essex initiative is now clearly something of a beacon for others interested in community urban farming; as Steve says, it's a considerable job just keeping up with all the requests for help and advice.[391] While it is part of a spreading movement, it doesn't itself show that this is how entire populations could be fed: projects like Steve Green's get communities engaged with growing food but they do not claim to meet all the food needs of the people who participate, let alone whole urban areas. Examples of this sort of use of urban spaces for agriculture do exist on a larger scale, but not within the West. To see how urban farming could work as an answer to problems of food production, we have to turn to Cuba and Venezuela.[392]

In Cuba, the turn to urban farming began in 1989, when the collapse of the Soviet bloc cut the country off from the source of

85% of its trade, including 57% of its food and 50% of its fertilizer and pesticides.[393] The popular response to the resulting food shortages was to start growing their own and urban farming sprang up wherever there was space for it. By 2011 there were nearly 400,000 urban farms, supplying 70% of the vegetables eaten in cities like Havana and as a result of this (and land redistribution supporting peasant agriculture), Cuba could boast the best food production performance in Latin America.[394] It is a particularly noteworthy story because this is sustainable agriculture, carried out perforce without organo-phosphate fertilizers, oil-dependent mechanisation or long supply chains. Venezuela adopted a Cuban-inspired urban agriculture programme in 2003, much expanded in 2011 in the 'Agro Ciudad' programme.

The extent to which the Cuban and Venezuelan experience of sustainable agriculture points the way for world food production to follow is contested. In 2009, for example, Dennis Avery of the right-wing Hudson Institute claimed that in Cuba, 'the organic success was all a lie, a great, gaudy, Communist-style Big Lie of the type that dictators behind the Iron Curtain routinely used throughout the Cold War to hornswoggle the Free World'[395] as supposedly 84% of Cuba's food was imported. Avery's arguments have been demolished, notably by the revelation that in 2003 Cuba was importing only 16% of its total food supplies,[396] but that it was made at all demonstrates the extent to which the verdict on Cuba's urban agriculture depends on what you think of the Cuban system.

If this is the case for Cuba, it is true in spades for Venezuela, where the standard of living under the Chavez and Maduro governments has become part of the right-wing offensive against them. Since 2012, Western media have been reporting severe shortages of basic goods like coffee, rice, toilet paper and cornflour in Venezuela.[397] The reports have been so persistent and apparently authoritative that even Westerners with their own contacts in Venezuela have found themselves believing them. Lee Brown, of the UK's Venezuela Solidarity Campaign, told me that when he led a student delegation to Venezuela in the summer of 2013, he'd felt it important for them to be ready for the toilet paper shortage. 'Don't worry about clothes' he told the students, 'just fill your bags with toilet paper, as much toilet paper as possible.' When they arrived and saw the piles of toilet paper for sale in every shop, they quietly unloaded their supplies next to them and tried not to mention it.[398]

The Venezuela Solidarity Campaign points out that the stories of shortages are often exaggerated or downright untrue and campaign supporters have posted their evidence that goods can in fact be found in shops in Caracas, as for example photographer Sean Hawkey found in 2013.[399] The Venezuelan government blames such shortages as do exist on speculators who, whether for economic or political reasons, buy up goods in a particular area and then sell them out of the country, particularly to Colombia. It was for this reason that they introduced an ID card for shoppers at government supermarkets in April 2014.[400]

Arguments over the toilet paper supplies, or lack of them, in Caracas might seem a world away from the question of the possibilities of urban agriculture, but there is a connection. Just as accusations of starvation for the poor in Cuba were used by Avery to discredit Cuba's sustainable agricultural production, so too is the idea of shortages in Venezuela used as an argument for how Chavez's agricultural policies have failed. Thus an analyst at 'one of the country's leading consulting firms' was quoted in *The Guardian* in September 2013 blaming food shortages at least in part on 'the decrease in agricultural production resulting from seized companies and land expropriations'.[401] It is perhaps not unreasonable to view this as a criticism of Chavez and Maduro from the right and there certainly seems an element of ideology in the statement that land redistribution axiomatically results in lost productivity. However, even attempting to strip out the most obviously right-wing political conclusions being drawn about Cuban and Venezuelan agriculture, there is a more moderate position which would point out that while both countries have indeed seen significant improvements in agricultural production and reductions in poverty and malnutrition in their poorest citizens, this is nevertheless when starting from an extremely low base. The urban agriculture initiatives of Havana and Caracas may well show how we could feed the world sustainably, this argument would have it, but only if we were all prepared to abandon Western standards of living and live as the poorest do in Havana and Caracas. Considering the possibilities for agriculture demonstrated by the Cuban and Venezuelan experience underlines that this is not merely a question of agriculture.

At a theoretical level, of course, we can look at the question of how we can produce enough food to feed everyone, without wrecking the climate, entirely aside from social questions. Studies of existing models, from Amish farming methods to urban collectives in Ontario, suggest that a shift to a more intensive but much more sustainable form of agriculture would greatly reduce the greenhouse gas emissions for which it was responsible. This in itself does not presuppose anything about how this production should be organised. Family-sized farms do not have to be farmed by families, nor would sustainable agriculture always have to be based on privately-owned small farms supplying the market. The fact that the most sustainable scale for agriculture is quite small does not preclude, for example, collective farms; it just shows that the idea that bigger is always better when it comes to farming is incorrect. If we get away from the idea that industrial production is always more productive than any other method, it does seem that it would probably be possible to feed everyone, even allowing for increasing living standards for the poorest. A recent consideration of the possibilities concluded:

> We do not yet know for sure whether a transition toward the sustainable intensification of agriculture, delivering greater benefits at the scale occurring in these projects, will result in enough food to meet the current food needs in developing countries, let alone the future needs after continued population growth and adoption of more urban and meat-rich diets. But what we are seeing is highly promising, especially for the poorest.[402]

The question however is whether that transition to sustainable agriculture could be achieved within the current system. The experience of Cuba and Venezuela after all has been of attempts to maintain food sovereignty while not treating food as just so many commodities and the recent history of both countries demonstrates how prepared the US in particular is to tolerate that. The lessons from both Cuba and Venezuela are that the extent to which this is possible within capitalism is at the very least open to question.

From food to the totality of capitalism

That it is possible to grow our food in a different way from that currently practised is a fact worth knowing, but it does not in itself solve the problems of food and climate change, because those current food production methods do not happen in a vacuum. Agriculture is part of the capitalist system, just as much as the supermarkets which sell us its ultimate products, and the way it is practised is increasingly determined by the needs of industrial capital, not only in the West but globally. The means by which large-scale capitalist enterprises have gained control over farming, which used to be carried out predominantly by small, independent producers, have been the introduction of things like fertilisers, pesticides and mechanisation, which have to be supplied to the farm from outside.[403] It is precisely these elements which would need to be reduced, if not eliminated, in the shift to a more sustainable agriculture. This is not a coincidence. The harm which capitalism does with agriculture arises from the way it treats the production of food – living things – as just another commodity to be traded and it is the production of standardised commodities which is enabled by these methods. It is also a setup which would prove difficult to dismantle under capitalism.

The awareness that it should be possible to produce sustainably enough food to feed the world, but only if we abandon the current capitalist methods, is not so much an answer to the problems of food and climate change, but a challenge. As discussed in the previous chapter, the damage and wastefulness caused by the food system under capitalism does not stop at the farm gate but encompasses all aspects of the system, from production through processing to distribution and disposal. A change in one part of the chain may not be useless, but would clearly have limited impact if the other links remain intact. This is not simply because that change would not be addressing the harm caused by other aspects of the food system, but because without a complete change in the system, no measures to ameliorate different aspects of it would be effective.

Capitalism as a system must be understood as a totality. We're encouraged to see different aspects of society as separate parts, which can be considered and changed without reference to the rest of society, but this is a fallacy of bourgeois thought. We can only understand the system if we see it as a total system, in which different elements are not so much static parts of a whole but elements of the process by which the system works.[404] This is not a philosophical abstraction, but an insight with very practical implications for how we can achieve systemic change. In particular, it exposes the limits of the sort of green campaigning which focuses on changes to individual behaviour, or in specific areas, on the basis that 'if everyone did X, then the effects would be enormous'.

As far as food is concerned, we are used to the idea that people should avoid supermarkets, air-freighted food, vegetables grown in polytunnels, things wrapped in layers of plastic and so on. It would therefore be easy to take the revelation that different farming methods could have the potential to produce the food the world needs without the current environmental harm and turn it into a similar programme for action. Everyone should get an allotment, we might conclude, or everyone should lobby their local council for more land to be set aside for urban agriculture. I am not arguing that these things would be actively counterproductive, but it is clear that they could have only a limited impact in the conditions existing under the current capitalist system. Land for allotments in major cities is limited and there are long waiting lists for plots. This is for the same reason that the giving over of large amounts of urban land for growing things is unlikely: the profit which could be made by using such land for development is considerable, whereas turning urban land over to agriculture would generate very little profit, if at all.

Even the common-sense argument that there can often be an oversupply of office blocks or luxury flats for short-term rent does not overcome this reality. We might recall how in the last property boom buy-to-let investors were snapping up city centre flats and then leaving them empty, so that they could make money by selling them on when property prices had risen some more. In capitalist terms, a building, even an empty one which no-one wants to rent, is a commodity which can be traded and on which finance can be leveraged. A plot on which to grow things has very little capital value, and the fact that the produce from the allotment might be more useful to individuals and society than more unused office space has nothing to do with it.

The notion that individuals can create demand for more sustainable food by changing their own eating and shopping habits falls foul of the same sorts of problems. To argue that people simply need to understand that eating processed food or takeaways is bad for them and bad for the planet, as often seems to happen in green discussion of food and eating habits, is to assume that people only eat this way out of stupidity or laziness. Because food has acquired such a wealth of moral significance, concluding that people eat differently from how we would like them to because they are bad people is in many ways the easiest option, but it does little to get us to a more sustainable food system. It is more useful to understand individual food choices in the context of the system in which they are made; a system in which, for example, the much-despised junk food is demonstrably cheaper per calorie than many of the supposedly 'better' options which working-class people in particular are hectored that they should eat.

Coupled with the fact that for many people what we might consider to be more sustainable eating patterns are impossibly expensive, is the related consideration that many simply do not have the equipment (cooking implements, decent kitchens, cookers) to prepare food themselves from scratch, nor the time to do so. If we decide that the problems of food and climate change are best solved by impelling individuals to change their behaviour, for this to be a mass shift in eating patterns we would need also to raise wages, decrease working hours and supply a considerable amount of capital investment to upgrade housing stocks. Individuals don't eat processed food because they want to destroy the planet, they eat it because it is the most convenient option, and that convenience matters not just to individuals, but to capitalism. If the cheap, quick food did not exist, the cost of the reproduction of labour, which has to be met if the workforce is to be maintained, would be much higher.

The point is that the way that food is produced and the way that we eat is an inextricable part of the capitalist system in which we live. It may well be possible for individuals to change their own lifestyles, but the extent to which they can do so, and the sort of changes which they can make, are themselves determined by their place in the system. Those changes don't subvert the system itself. Even individuals deciding to drop out of the system altogether does not add up to systemic change, since it isn't a change which more than a very few people would be able to make. This isn't because only a few people have the willpower, moral courage or hardihood to live without mains electricity, iPods or McDonald's, but simply because, within the totality of capitalism, individuals who decide to live without the trappings of it are able to do so because of the existence of the system. It would not, for example, be possible to be a freegan without the functioning system throwing enough food away for other people to be able to live on.

The idea that we can prefigure the change we want to see in society – start living now in the way that we would like everyone to be able to live – is a powerful one. Exercises in constructing a different sort of society, from Climate Camp to the Occupy movement, demonstrate that there could be alternatives to the capitalist system and this is often at least part of the thinking behind community farming projects. As the Director of the Portland-based Urban Farm Collective said in 2013, for example, the long-term aim for the project was to get 'a critical mass of a collective that's outside the cash-for-goods economy system', teaching people 'to be self-reliant in regard to the food they eat... I feel like there's a shift happening. People are getting away from consumerism.'[405] Convincing people that there should be an option other than capitalism is clearly worthwhile but it is not the same as defeating capitalism itself. If we want a sustainable food system, the answers do not lie in changing the way that we eat or think about what we eat, still less in changing the way that despised others (working-class, poor, fat people) eat. What must be done is to address the system which determines the way that our food is produced, chosen and wasted.

The discovery of food as an important cause of greenhouse gas emissions has been an exercise in reducing the issue of climate change to atomised, individual choices: to eat meat, to be fat, even to exist at all. This carries with it the notion that some groups are more responsible for making poor choices than others, and neatly shifts the responsibility for climate change, within a model which sees it arising from overconsumption, from the luxury consumption of the genuinely rich to that of those on the receiving end of the increasing income disparities in Western countries. It is a rightward shift: it is difficult to stand in solidarity with working-class people if you think that one of the most serious issues facing society is actually all their fault. However, the argument over whether or not it is possible to feed everyone on the planet without destroying the planet does not have to be reduced to a seesaw between the Malthusian view and the industrial technofix. Ultimately, it is the understanding of capitalism as a historical system, with a beginning and an end, which breaks through this duality. If capitalism is not an eternal, natural state, then we should be able to separate the effects of capitalism, which are specific to this system, from the effects of nine billion people living on the planet which would be inescapable under any system.

This matters because what the food system demonstrates is not that climate change is caused by individual actions, but that it is an epiphenomenon of capitalism. The food system uses the energy it does and causes the emissions it does because it is effectively set up to do so, because those energy inputs and outputs represent the profit which is the driver of every part of the whole system, not just with reference to food. This understanding indicates that regardless of what we choose to eat, the environmental impact is likely to be dire, because the options to choose differently won't exist unless they are profitable, and therefore destructive. Individuals are encouraged to replace red meat in particular with a vegetarian option, but processed vegetarian food would have its own environmental impacts, and a vast increase in soya production in particular would be especially unhelpful in climate terms.

This means therefore that if we care about the effects of our food system on the climate, and about our ability to feed the world, we cannot fight specifically on the food issue. Indeed, to make food and climate change into a discrete issue, related to but separate from other issues of climate change, would be to perpetuate the problem which needs to be solved. The approach to climate change which specifies and divides the different contributors of greenhouse gases as if they were separate singles fighting it out in the Top Ten is a way of concealing the relationship of climate change to the entire system, whereas dealing with it necessitates seeing it as a whole.

Practical action: the British climate movement today[406]

It is clear that the problems of food and climate change won't be resolved by campaigning on food as an individual issue. But a prescription that anyone concerned about food's contribution to climate change should simply involve themselves generally in the movement is not uncontentious, given the position and recent criticisms of the British climate movement.

It is possible to view the protests accompanying the international climate summit in Copenhagen at the end of 2009 as a high point for the climate movement, from which the subsequent years have seen a steep decline. The numbers demonstrating are themselves indicative of this: the protest at Copenhagen itself was made up of at least 100,000 people, while Stop Climate Chaos' 'The Wave' demonstration in London culminated in 50,000 people encircling the Houses of Parliament in the largest ever climate protest in the UK. In contrast, subsequent protests have often barely achieved the one thousand mark (although the People's Climate March in September 2014, as this book was in production, saw an impressive 30,000 people on the streets of London). If this has implied a decline in the movement, it is not restricted to demonstrating alone. Climate Camp, arguably one the most visible parts of the movement, decided to wind itself up to allow its members to concentrate on other things, suggestive of a sense that climate change was no longer the key issue for predominantly young, radical activists. To cap it all, a global poll in February 2013 concluded that public engagement with climate change worldwide was at its lowest for 20 years,[407] suggesting that the weakness of the climate movement since Copenhagen may have had an effect on public consciousness of the issue.

In one view, the movement is simply the victim of unfavourable objective circumstances. The furore over 'Climategate', the hacked emails from climate researchers at the University of East Anglia, along with the contemporaneous accusation of mistakes in IPCC reports, empowered a new wave of climate change denial, while the failure of world leaders to come up with a meaningful deal at Copenhagen dealt a significant blow to the idea of climate change as a solvable problem. Even the weather played a part in the creation of these difficult times for the green movement. Popular confusion between weather and climate may have helped keep climate change in the news when 'the hottest X since records began' headlines gave journalists an opportunity to report on it, but with the downside that particularly cold winters or wet summers can then be taken as evidence that climate change is not happening.[408] Alongside these difficulties, it also could be argued that the economic crisis has pushed climate change down the political agenda, although this does not explain why the anti-war movement did not have a similar effect on climate change campaigning.

That Climategate and Copenhagen were a difficult double whammy for the green movement is fairly uncontroversial, but there is another argument which sees climate change campaigns as the authors of their own misfortunes, actually damaging the agenda they are seeking to promote. This is not solely a post-Copenhagen contention: Ted Nordhaus and Michael Shellenberger of The Breakthrough Institute put their case that the green movement was neither right for nor up to the task of combating climate change in 2004, followed in 2007 by their book *Break Through. Why We Can't Leave Saving The Planet To Environmentalists*.[409] It has however been given extra impetus by the sense of crisis after the failure of the Copenhagen summit. Since early 2010 criticisms have been levelled at climate change campaigners by figures who could be seen as having come from within the movement, and who were certainly widely respected in it, such as Mark Lynas and George Monbiot.

The issues which Lynas and Monbiot have expressed with the green movement are various, but at base they come down to the same as are revealed by the presentation of food as particular important for the climate. The attacks by Lynas and Monbiot, among others, on the climate movement were traumatic, given these writers' status for many campaigners, but they are also important because of what they reveal about the position of the movement. The debate which they have opened up is not about tactics or the failings of specific groups, however much it may appear to be only about specific issues such as nuclear power. The key question is whether the response to the failures of climate summits should be to understand the way in which the system itself stands in the way of action on climate change, or simply to relinquish the idea that any significant change is possible.

The most obvious point of contention between Lynas and Monbiot on the one hand and many greens on the other is the issue of nuclear power. For Lynas, nuclear power is a clear necessity, so much so that he states in his latest book, *The God Species*, that 'anyone who still marches against nuclear today... is in my view just as bad for the climate as textbook eco-villains like the big oil companies'.[410] Monbiot is less unequivocal than

Lynas in his nuclear position, but he increasingly regards nuclear power as a least-worst, necessary replacement for coal.[411] Like Lynas, he appears to see green opposition to nuclear as not only mistaken but in rather bad faith, commenting for example in a reply to Jonathan Porritt in August 2011 that we should ask ourselves if the aim is to stop climate change or to get maximum take-up of renewables, as it sometimes seems that greens prioritise the latter over the former.[412] Both Lynas and Monbiot therefore present a picture of green opposition to nuclear power which is ideological rather than arising from an understanding of the nature of nuclear power.

For both Lynas and Monbiot, the nuclear argument became more urgent to win as a result of the decision of the German government among others to abandon nuclear power following the explosion at the Fukushima nuclear plant in Japan in 2011. They both argue that Fukushima should not be seen as a confirmation of anti-nuclear arguments but the final contradiction of them. It shows, they think, that even in the worst circumstances imaginable, the consequences of the explosion were actually fairly limited, particularly, as Monbiot points out, in comparison to the likely death toll from unchecked climate change.[413] Since the effect of the disaster on the long-term health of the surrounding population will obviously not be clear for many years, this confidence in its limited effects may be rather premature. However, the point is not Lynas and Monbiot's support for nuclear power *per se* but what it shows about their views of the problems of the green movement.

Lynas and Monbiot are clearly coming from different places politically, and both differ again from the position of the Breakthrough Institute, but they are united by what seems to be an appeal to pragmatism. In different ways, all these arguments present the idea that one of the major problems with the green movement is that it has become too ideological, and that ideological nature is preventing it from simply embracing what might work to deal with climate change. This is not restricted to the nuclear issue; for both Lynas and the Breakthrough Institute, the root of the problem lies in the fundamentals of what they characterise as green thinking.

The central opposition for these authors is between green desires for untouched nature, secured from human influence, and those who, like them, appreciate the primary importance of human society. It is true that the idea of a pristine natural world is impossible. The notion itself results from our alienation from nature under capitalism, and a failure to appreciate the dialectic between human society and the natural world, shaping and being shaped by each other. Lynas and the Breakthrough Institute are not however asserting the importance of understanding this dialectic, but making the extremely undialectical argument that what matters is the irreversible status of man's domination over nature.

That real domination is meant is shown by the metaphors used to describe it: for Lynas, we need to embrace our role as the conquerors of the world and realise that 'the first responsibility of a conquering army is always to govern',[414] while for Nordhaus and Shellenberger, our position *vis a vis* nature is even more lofty, as they quote US environmentalist Stewart Brand, 'We are as gods and might as well get used to it'.[415] Lynas' expression of human dominance of nature in military terms is telling in the context of his argument that greens have allowed a straightforward message about climate change to become contaminated by other political agendas. Given the involvement of many greens in campaigning against imperialist wars in Afghanistan and Iraq in recent years, the characterisation of humans in general as a conquering army comes across as a very definite statement of position; a challenge to the totality of what the green movement might be thought to stand for.

The idea that all environmentalists pursue the separation of nature from humans is of course highly debatable. Reviews of the first edition of *Break Through* pointed out that the authors had set up a straw-man environmental movement against which to argue. Nordhaus and Shellenberger responded in the second edition with the unanswerable claim that any objections only prove them more correct: 'Perhaps the reason these environment exemplars [cited in the book] seem so alien is because modern green readers have experienced a shift in their own values, one that has occurred gradually over the past decade. It may be that we are all now, to one degree or another, post-environmentalists.'[416] Clearly, these authors do believe their characterisation of the green movement as technophobic fundamentalists is a fair one, but it's also the case that framing the argument in these terms enables them to start from the implication that the green position is constituted by opposition to any technical innovation to deal with climate change.

The debate is thus portrayed as between on one side a green view which sees energy-saving austerity and cuts in living standards as the only way to reduce emissions, and on the other the sane, balanced thinkers who perceive that modern technology can do the job on emissions while enabling us to keep the lights on. Lynas sets out the position effectively: 'Global warming is not about overconsumption, morality, ideology or capitalism. It is largely the result of human beings generating energy by burning hydrocarbons and coal. It is, in other words, a technical problem, and it is therefore amendable to a largely technical solution, albeit one driven by politics.'[417] It is an eloquent statement, but is only achieved by conflating green suspicion of technofixes with technology, a conflation which reveals rather neatly the ideology beneath Lynas' claims to a non-ideological position.

The notion that greens in general are opposed to using technology to reduce emissions is at the very least difficult to sustain. Opposition to the concept of renewable energy, for example, is not a mainstream green position, nor is opposition to the development and improvement of these sorts of technologies.

The problem with the technofix is not the use of technology itself, as opposed to the return to the pre-technology age for which greens are supposedly nostalgic, but the attempt to use technology as an alternative to making changes in the way the current system works.

Technofixes, in other words, are technological measures designed to permit the continuance of the system which created the problem in the first place. They maintain the *status quo* and Lynas appears to embrace them for precisely that reason. Lynas' argument is essentially that solutions to climate change can only be found within the current system, and they can therefore only be technical ones. This is presented as a non-ideological position but it is of course intensely ideological: arguing that no solutions for climate change which call for system change are legitimate is either an argument that the current system is the best possible, or, at the very least, that attempts to change it are doomed to failure.

Given this basic assumption that only solutions within the current system are acceptable, it is not surprising that Lynas views greens who are critical of the system as actively unhelpful as well as misguided. They drive away those who have a stake in the current system, who need only to be convinced that dealing with climate change does not mean consorting with hairy lefties. This is perilously close to the argument put by Anthony Giddens in his *Politics of Climate Change*,[418] which was that governments and market leaders have failed to act on climate change until now simply because to do so would entail associating themselves with such dreadful people as greens.

This is an important argument because Lynas and Giddens have to explain why, for all the eminent solvability of climate change within the current system, it has not, in fact, been solved. The technofixes may be easily available, but evidence for their large-scale imminent deployment is sadly lacking. If this failure cannot be laid at the door of counterproductive green campaigners, the only option is to conclude that the system is not able to respond to climate change by introducing measures to combat it directly, despite persuasive arguments that preventing catastrophic, irreversible warming will be far less costly than allowing it to happen and accepting the costs of attempting to live with it.

This has been the conclusion for example of a number of climate scientists who produced the 2010 Hartwell Paper, effectively arguing that actions to address climate change for the sake of addressing climate change are impossible, and that the only way forward is to encourage changes which provide direct economic benefits, and climate benefits as a side effect.[419] The Hartwell Paper was written explicitly as a response to the 'crash of 2009': the twin effects of Climategate and Copenhagen, which for the authors meant that existing approaches to climate change were completely discredited. While Lynas presents his arguments within a more positive framework, it is difficult not to see *The God Species* in a similar light, as a product of the perceived defeat of the green movement in 2009.

George Monbiot, while writing from a more generally left-wing perspective than Lynas, seems to have a similar take on the movement, in that the perspective now is one following a significant setback. Monbiot's adoption of a broadly pro-nuclear power position can be seen as a study in his changing view of the green movement. He has come to hold a position that it has been damaging its own cause by refusing to adopt a pragmatic view of current possibilities. This is because, while it is possible to trace in Monbiot's recent writing a move towards more enthusiastic support for nuclear power, it is clear that he remains far from the wholehearted nuclear enthusiast that Lynas appears to be. For Monbiot, nuclear power remains a far from ideal solution, but is the solution which is both available now, and is significantly better in climate terms than fossil fuel power stations.

That this is a capitulation to the *status quo* is demonstrated by the fact that Monbiot does not dispute that it would be theoretically possible to decarbonise energy generation, noting that Zero Carbon Britain and others have shown how this could be done in principle through a combination of renewable energy generation and energy efficiency.[420] The barriers to a large-scale shift to renewable energy are not physical but political: the additional costs of managing the grid with a significant proportion of electricity from renewable sources, the time it would take to construct the new infrastructure, and local objections to new renewable installations, particularly to wind farms and to the new power lines they would require.[421]

It is interesting that for Monbiot protestors against wind farms are a political reality whose views have to be accommodated, even to the extent of abandoning the idea of a decarbonised energy system derived from renewable sources, yet anti-nuclear protestors simply have to accept the error of their ways and stop their opposition to this best available answer to climate change. Why both groups of protestors do not have equal weight is not clear. Even if any suggestion of a preference for the anti-wind farm over the anti-nuclear protestors were to be dismissed, an argument that anti-wind farm protestors have more political weight than anti-nuclear ones is difficult to sustain. While wind farms are a country rather than an urban issue, this does not automatically make them more influential even under a Tory government, and it is worth noting that the areas with the most anti-wind farm campaigns are not the traditional Tory shires of south-east England, but Scotland, Wales, northern England and Devon and Cornwall.[422] Compared with fears expressed by both Monbiot and Lynas that in the wake of Fukushima anti-nuclear campaigns could derail a new generation of nuclear power stations, this does not appear to add up to a uniquely powerful anti-wind lobby, so why they should be accommodated while anti-nuclear protestors are dismissed remains unexplained.

The other arguments in favour of nuclear as opposed to

renewable energy also work only with the assumption that the system, and therefore the rules within which the different options for low-carbon energy generation are considered, cannot be changed. This is the assumption that increased generation costs could not be met with increased government subsidies, and would therefore always end up with higher bills for consumers. It similarly underlies calculations of how long the creation of a renewable infrastructure would take. Statements of the time needed for large-scale construction projects can appear to be describing immutable natural laws but, of course, they aren't; they are estimates of how long projects would take with a given level of labour and investment. This level could be changed, and could be changed more markedly if the profit requirements of private companies did not have to be taken into account. Not everything has to take several years; even in a capitalist economy, changes can be made remarkably quickly if there is sufficient motivation, as the US economy's 'turn on a dime' when they entered the Second World War showed.

However, since Monbiot effectively rules out any such dramatic shifts in the current economy, the task of shifting energy generation to renewables appears an impossible one. Unlike Lynas or the Breakthrough Institute, Monbiot remains unconvinced by the possibility of responding to climate change within the current system, yet for him no solution which looks beyond the current system is possible. It is small wonder that he sees the green movement as lacking in answers, given the unanswerable conundrum this reasoning sets up, nor that his writing on energy generation from 2010 on has become steadily more pessimistic. This sense of pessimism is very much in keeping with the idea that Copenhagen and Climategate added up to a defeat for action on climate change, but it is clearly related here to the assumption that a practical response means not looking beyond the *status quo*.

This, then, is why advice simply to get involved in campaigning for action on climate change is not straightforward.

Where once it was possible to join the annual march for governments to do something – anything – to reduce emissions, now it is necessary to decide whether we support market mechanisms to encourage technofixes within the current system, or whether it is systemic change for which we need to fight. It is clear that some of the reorientation of groups within the climate movement has happened around this requirement. It has been suggested that part of the reason for Climate Camp's decision to release its participants for other projects was the difficulty in moving from an understanding of themselves as activists against climate change specifically to more general anti-capitalism within the Climate Camp framework. Climate Camp could not itself take on the entire system, but there seems to have been an increasing sense that concentrating only on the climate change aspect of the wider problem was not enough.[423]

This is not to say that campaigning on climate change as a single issue is unimportant. As suggested, the depressing conclusion that the public worldwide are now less concerned about climate change than at any time in the last twenty years suggests that without a strong campaign keeping the issue in the news and in the public eye, it can be allowed to slip quietly down governmental agendas, meaning that we will lose the chance of even minor action to reduce emissions. It is also clear that as oil-dependent industries come up with more and more damaging ways of extracting oil and gas, such as fracking shale for natural gas, there will be campaigns to resist these; campaigns which will be greatly strengthened if they can situate themselves within national campaigns for action on climate change. However, an understanding that the only way to address climate change is to see it as a creation of the capitalist system in which we live, demands that campaigns to address this locate themselves within struggles against other aspects of the system.

Concluding that the problem of food production under capitalism is a problem of capitalism, not of those who eat the food, does not mean that there is no action that we can take. This is not an argument for sitting back until the revolution. As noted above, the arguments for addressing climate change through addressing the diets of predominantly poor people in the West are the same arguments used to justify government austerity. We have caused the problem with our own profligacy – our lack of moral restraint, as Malthus might have chimed in – and if we are now suffering the consequential cuts in public services and living standards, perhaps it will teach us to be more careful with the credit card in future.

It is sometimes noted by writers on growth and its relationship to climate change that recessions can be effective at reducing carbon emissions: Russia in the 1990s, for example, had relatively low emissions not because it had been successful in cleaning up its fairly appalling environmental record but because of its economic collapse. Supposedly environmental justifications for cuts have not figured highly in governmental propaganda up until now, but it is possible that this is only because climate change itself is not high enough on most people's agendas to make it a valuable argument. Since views on climate change do appear to be as changeable as the weather, this could also change. It is entirely possible that we will see a government minister defending squeezing food budgets through benefit cuts with the argument that we all have to eat less anyway for the sake of the climate and the starving in the Third World. If we don't want to see this co-option but do want a different way of living and to save the planet, we need to be fighting for action on climate change as part of the wider fight against austerity.

If we understand climate change as not a technical problem but an integral part of capitalism then joining these campaigns *is* fighting for climate action, and it is concentrating the fight where we have the greatest chance of building a genuinely mass movement. There is also considerable synergy between climate change issues and the fight against austerity. As the campaign for climate jobs has demonstrated, pointing out the sense in meeting the economic crisis by creating jobs and a green infrastructure can get wide acceptance as a Keynesian alternative to ordinary people paying for the bankers' crisis.

The conclusion that fighting for a better food system involves fighting on issues which do not appear to have anything to do with food is only contradictory if we lack an understanding of how different parts of the system form aspects of the totality, rather than existing as discrete problems with their own solutions. If we see that it is the whole system which we need to change, it makes sense to concentrate on issues on which we have the best chance against it.

What we would eat in Utopia

This leaves aside the question of what our food system could look like under a different economic system, which might seem a rather crucial omission. How, after all, can we fight for a better system if we don't know what we are fighting for? It is however well-nigh impossible to talk about how we might eat after the revolution without appearing to suggest that this is how revolutionaries should try to eat now, with all the implied guilt if they fail to do so. Food is such a personal issue, and in our society, one laden with such moral significance, that discussions of how we might eat better can carry more implications of prescriptions for immediate, individual action than would be the case for other areas of society.

This is not to say that it has not been attempted, as food has a place in a number of imagined post-revolutionary Utopias. In William Morris' *News from Nowhere*, for example, food is simple but plentiful and certainly not lacking in variety, as this description of an everyday breakfast shows: 'simple enough, but most delicately cooked, and set on the table with much daintiness. The bread was particularly good, and was of several different kinds, from the big, rather close, dark-coloured, sweet-tasting farmhouse loaf, which was most to my liking, to the thin pipe-stems of wheaten crust, such as I have eaten in Turin.'[424] While this breakfast seems to be an individual or at least a household affair, dinner in London later proves to be more communal, heralded by a city-wide dinner bell and served in a large dining hall. Like the breakfast, though, it is simple, with 'no excess either of quantity or of gourmandise' but 'excellent of its kind'.[425] This is clearly less of a thoughtful model of how a post-revolutionary society might actually eat than it is a reaction against what Morris perceived to be the two models of nineteenth-century eating: the adulterated, low-quality food available to the poor versus the over-embellished, fashionable dinners of the wealthy. He makes the point that in his utopian dinner 'there was a total absence of what the nineteenth century calls "comfort" – that is stuffy inconvenience; so that… I had never eaten my dinner so pleasantly before'.[426]

The arrangement imagined by Morris, of communal main meals combined with smaller meals in the household, is also followed by Morris' contemporary, Edward Bellamy, in his *Looking Backwards*, which imagines an inhabitant of 1887 Boston who sleeps 'in a trance' until a post-revolutionary year 2000. In twenty-first-century Boston, apparently, 'the two minor meals of the day are usually taken at home, as not worth the trouble of going out; but it is general to go out to dine'.[427] As it turns out, this experience is less communal than the dining imagined by Morris, as each household has their own private room in the hall. Food is admitted to be important in this society: the narrator is told that 'There is actually nothing which our people take more interest in than the perfection of the catering and cooking done for them, and I admit that we are a little vain of the success that has been attained by this branch of the service'.[428] Beyond that, Bellamy did not consider what sort of food might be eaten; his point is simply that, in his Utopia, everyone can finally eat what they want.

A more modern take on eating after the revolution comes from the science-fiction writer Ursula Le Guin. Le Guin's best-known post-revolutionary society is in her novel *The Dispossessed*, which portrays a socialist community about 200 years after establishment. In many ways, the consumption patterns of the people of Anarres is similar to the vision of a sustainable world painted by many modern writers on overconsumption. They have enough to eat (except when there is a famine), consumed in communal dining rooms, but no more, and treats are severely limited. When one of the characters wants to hold a celebration, for example, he saves up his special beverages ration for ten days to provide a litre bottle of fruit juice.[429] So strict are their rules to control wasteful consumption of all material resources that private property is disallowed, even down to the most insignificant personal item. When, in a moment of emotional crisis, the main character needs to dry his tears, his small daughter offers, conscientiously, 'you can share the handkerchief I use'.[430]

This portrayal of unmaterialistic scarcity might resemble some environmental calls for restrained consumption, but it shouldn't be taken as an attempt to imagine what a post-revolutionary world might look like. The point in *The Dispossessed* is that the people of Anarres are not alone, but live in the shadow of a wealthy neighbouring planet, Urras, from which their ancestors came as settlers. The Anarresti were allowed to leave Urras and set up their socialist colony because exiling them to a planet so barren that it could barely support life was the easiest way of dealing with them, but their fate is related to that of their sister planet. The Anarresti are effectively practising socialism in one country, with all the difficulties that implies, not living in a post-revolutionary Utopia.

Le Guin's really Utopian novel is not *The Dispossessed* but the lesser-known *Always Coming Home*, in which she portrays a non-class society of villages in a far-future northern California. These people, the Kesh, are subsistence farmers and hunters, using little technology, but they produce an abundance of varied food. 'The Kesh took or found and raised or grew food wherever convenient, and cooked and ate it with interest, respect and pleasure.'[431] Le Guin imagined that there would be some social disapproval for overconsumption, just as hoarding personal possessions was also frowned upon: 'heavy eating was considered embarrassing and gorging shameful'. The Kesh, however, did not let this prohibition stop them from eating when they wanted, as 'greed could be satisfied more or less invisibly by casual but persistent snacking'. Le Guin ends this passage with what is possibly the most optimistic summary of what we might eat in Utopia that I have ever come across: 'The Kesh were not a thin people'.[432]

Always Coming Home is particularly interesting because it represents an attempt to imagine what a non-class society might look like, as opposed to using the notion of a Utopia to make explicit points about what is wrong with present-day society (not that the latter is an illegitimate exercise). In general, however, literary versions of post-revolutionary societies do not get us very far in considering what the ideal food consumption pattern might look like. *News from Nowhere* and *Looking Backwards* might tell us that Edward Bellamy thought everyone should be able to eat like the wealthy of 1880s Boston, while William Morris thought the eating habits of the rich should be abandoned for a simpler diet, but this says little about how a post-revolutionary world would eat.

It is in fact not only impossible but counterproductive for us to prescribe what a different society after capitalism might look like. We are after all the products of capitalism, with the alienation caused by capitalist society, and we are looking towards a society in which people will no longer be distorted by capitalism and will therefore no longer be alienated. We can comprehend that an equal society in which we are not reduced to components of labour power would be a society which could see a great release of human potential, but we can't set down now what the fruits of that release of potential would be. Indeed, it would not be helpful for us to detail what we thought it should do, as our ideas from within capitalism should not be used as prescriptions for what society free from capitalism would prioritise.

What can usefully be said is that the strain put on the ecosystem by the current food system is a strain created by the destructiveness of capitalism, not by the inherent weight of the global population or still less inherent human nature. We can't know for sure how easy it would be to feed the world without the current capitalist system in the way, but we don't have to assume that it would need to start from a position of scarcity. This is not a prediction that after the revolution, everyone will have burgers and chips for every meal, although it might not be as problematic as might be thought if they did. Speculatively, if lack of time and money weren't issues, people might well choose to eat less meat-heavy diets than the modern Western model, and might well eat a greater range of fresh foods as it would be easier and pleasanter to do so. This is however still to think about food and eating through the lens of modern capitalist society, which isn't necessarily applicable to a post-capitalist world. A revolution would change not just the constraints on food production but the entire way in which production and consumption were organised. The Russian revolution after all socialised not just production but also housework, with the provision of workers' canteens so that food preparation did not have to be done in the home.

This is not, however, the point. A turn towards food as a key issue for climate change is a turn inward, towards an understanding of climate change that sees it as essentially unconnected to the system in any meaningful way. In this version, climate change is created by the cumulative effect of freely-assumed individual behaviours, and the answer therefore lies in reaching those individuals, one by one. In the context of the conclusion drawn by figures like George Monbiot that it is not possible to make changes to the system to address climate change, we can see how a turn back to an individual model of response to climate change might be appealing. If you can't change the system, maybe you can change people? The problem with this is that it is not individual behaviours but the system within which they are adopted which is the source of climate change and any number of dinners switched from processed beef to processed soya will not change that. The answer to climate change issues is not to retreat back to the belief that our moral food choices make a difference, but to understand that if we care about climate change, we have to join the fight for a more just system, which, through our involvement in that struggle, must also be a more sustainable one.

Concluding that it makes little difference what individuals choose to eat or not eat is not the same as arguing that what individuals do doesn't matter. But the place for climate change campaigning is as a central part of the fight against the depredations of the capitalist system, not standing on the sidelines analysing what either side is eating.

Notes

1. *Weathercocks and Signposts. The environmental movement at a crossroads*, (World Wildlife Fund, April 2008).
2. http://www.ciwf.org.uk/news/factory_farming/lecture_calls_for_dietary_change.aspx
3. *The Observer*, 7th September 2008, http://www.guardian.co.uk/environment/2008/sep/07/food.foodanddrink
4. *The Guardian*, 30th September 2008, http://www.guardian.co.uk/environment/2008/sep/30/food.ethicalliving
5. *The Telegraph*, 15th October 2008, http://www.telegraph.co.uk/earth/earthnews/3353377/Government-advisor-eat-less-meat-to-tackle-climate-change.html
6. George Monbiot, *Heat. How to Stop the Planet Burning*, (Allen Lane, London 2006).
7. Jonathan Neale, *Stop Global Warming. Change the World*, (Bookmarks Publications, London 2008).
8. Chris Goodall, *How to Live a Low-Carbon Life. The Individual's Guide to Stopping Climate Change*, (Earthscan, London-Sterling VA 2007), pp.230-46.
9. Pat Thomas, *Stuffed. Positive Action to Prevent a Global Food Crisis*, (Soil Association 2010), p.143.
10. http://news.bbc.co.uk/1/hi/health/7404268.stm, 16th May 2008.
11. http://www.guardian.co.uk/environment/2009/oct/26/palm-oil-initiative-carbon-emissions
12. 'Turn veggie to save planet, says Sir Paul', *The Independent*, 29th November 2008.
13. Raj Patel, *Stuffed and Starved. Markets, Power and the Hidden Battle for the World Food System*, (Portobello Books, London 2007), p.1.
14. See for example A. Haroon Akram-Lodhi, *Hungry for Change. Farmers, Food Justice and the Agrarian Question*,

(Fernwood Publishing, Halifax and Winnipeg 2013), p.4.

15. See for example Colin Tudge, *Feeding People is Easy*, (Pari, Italy 2007), p.10; and *Achieving Food Security in the Face of Climate Change*, Commission on Food Security and Climate Change (2011), p.3.
16. Thomas Princen, 'Consumption and Environment: Some Conceptual Issues', *Ecological Economics* 31 (1999), pp.347-63, p.348.
17. *Achieving Food Security*, p.4.
18. D Pimental et al., 'Reducing Energy Inputs in the US Food System', *Human Ecology* 36, no.4 (August 2008), pp.459-71, reviewed at www.esciencenews.com/articles/2008/07/23 /why.eating.less.can.help.environment.
19. The US average food availability per head is sometimes given as 3,774kcal, for example by Gideon Eshel and Pamela Martin ('Diet, Energy and Global Warming', *Earth Interactions* 10, (March 2006), pp.1-17, p.2), to which Pimental's figure of 3,747kcal seems remarkably similar.
20. Haroon Akram-Lodhi, *Hungry for Change*, pp.4-26.
21. See for example Ian Roberts with Phil Edwards, *The Energy Glut. Climate Change and the Politics of Fatness*, (Zed Books, London and New York 2010), esp. pp.48-65.
22. C Bouchard and S N Blair, 'Introductory comments for the consensus on physical activity and obesity', *Medicine and Science in Sports and Exercise* 31, 11, section 4, pp.98-501, p.499.
23. Tara Garnett, *Cooking up a Storm. Food, Greenhouse Gas Emissions and our Changing Climate*, (Food Climate Research Network, Centre for Environmental Strategy, University of Surrey, 2008), p.111.
24. Roberts with Edwards, *Energy Glut*, pp.136-8.
25. Akram-Lodhi, *Hungry for Change*, p.167.
26. Jamie O'Neill, 'Fat Bastards', *Sacramento News Review*, 28th June 2007.

27. http://news.bbc.co.uk/1/hi/health/7404268.stm, 16th May 2008.
28. http://obesitytimebomb.blogspot.co.uk/
29. Charlotte Cooper, personal interview, 10th October 2013.
30. Nicole Pontes, personal communication, 25th November 2013.
31. http://www.americasquarterly.org/tackling-brazils-obesity-problem, 27th June 2012.
32. *Weighty Matters. The London Findings of the National Child Measurement Programme 2006-2008,* London Health Observatory, (May 2009), p.12.
33. *Foresight - Tackling Obesities: Future Choice - Project Report,* Government Office for Science (2007), available at www.foresight.gov.uk, p.30.
34. www.msnbc.msn.com/id/27697364/, 17th November 2008. Huntingdon was number 1 in the fat city league in 2008 and had slipped to number 3 by 2012.
35. Actually I did this twice, in June 2009 and April 2012. I did not find any meaningful change in the representation of obesity in the intervening period – the main difference was that the news pieces were more likely to be illustrated with video in 2012 than they were in 2009, but in some cases this was a montage of headless working-class fatties.
36. See Nick Davies, *Flat Earth News. An Award-winning Reporter Exposes Falsehood, Distortion and Propaganda in the Global Media,* (Chatto & Windus, London 2008), pp.69-73 on what he dubbed 'churnalism' created in part by the demand for speed.
37. *Healthy Weight, Healthy Lives. A Cross-Government Strategy for England,* Cross-Government Obesity Unit, (January 2008).
38. Ibid., p.vii.
39. Ibid., p.xvi.
40. See www.dwp.gov.uk/benefit-thieves/local-authorities/

41. *Foresight – Tackling Obesities,* p.30.
42. Karl Marx, *Grundrisse. Foundations of the Critique of Political Economy (Rough Draft),* (1939) trans. Martin Nicolaus, (Penguin/New Left Review, London 1973), p.285.
43. Paul Campos, *The Obesity Myth. Why America's Obsession with Weight is Hazardous to your Health,* (Gotham Books, New York 2004), pp.57-68. The article discussed is Greg Critser, 'Let them eat fat: The heavy truths about American obesity', *Harper's,* (March 2000).
44. Peter Dauvergne, *The Shadows of Consumption. Consequences for the Global Environment,* (MIT, Cambridge, Mass. and London 2008), p.14.
45. Mark Gold, *The Global Benefits of Eating Less Meat,* (Compassion in World Farming Trust 2004), p.7.
46. Ibid., p.6.
47. Thomas, *Stuffed,* p.152
48. *Global Benefits of Eating Less Meat,* p.4.
49. Campos, *Obesity Myth,* p.3.
50. *Foresight - Tackling Obesities,* p.72.
51. http://www.cspinet.org/new/pdf/obesity-letter-obama.pdf, 22nd June 2009.
52. *Foresight,* pp.17-18. For a discussion of the connection of climate change and obesity in this report, see Rachel White, 'Undesirable Consequences? Resignifying Discursive Constructions of Fatness in the Obesity "Epidemic"', Corinna Tomrley and Ann Kaloski-Naylor (eds.), *Fat Studies in the UK,* (Raw Nerve Books, York 2009), pp.69-81.
53. *Food Matters. Towards a Strategy for the 21st Century,* The Strategy Unit, (2008), pp.15-16.
54. Phil Edwards and Ian Roberts, 'Transport policy is food policy', *The Lancet,* vol 371, no.9639, (17th May 2008), p.1661.
55. See for example Axel Michaelowa and Björn Dransfield, 'Greenhouse gas benefits of fighting obesity', *Ecological Economics* 66, (2008), pp.298-308; and Phil Edwards and Ian

Roberts, 'Population Adiposity and Climate Change', *International Journal of Epidemiology*, (2009), pp.1-4.

56. Michaelowa and Dransfeld, 'Greenhouse gas benefits of fighting obesity', p.299.
57. Eshel and Martin, 'Diet, energy and global warming', p.3.
58. Michaelowa and Dransfeld, 'Greenhouse gas benefits of fighting obesity', p.300.
59. For a review of recent studies and discussion of this point, see Michael Gard and Jan Wright, *The Obesity Epidemic. Science, Morality and Ideology*, (Routledge, London and New York 2005), pp.114-7.
60. Roberto P Trevino et al., 'Diabetes risk, low fitness and energy insufficiency levels among children from poor families', *Journal of the American Dietetic Association*, 108 no.11, (November 2008), pp.1846-1853, p.1849.
61. Gard and Wright, *The Obesity Epidemic*, p.45.
62. M. Berners-Lee, C. Hoolohan, H. Cammack and C.N.Hewitt, 'The relative greenhouse gas impacts of realistic dietary choices', *Energy Policy* 43, (April 2012), pp.184-90, p.187.
63. Michaelowa and Dransfeld admit that their calculations of the climate effects of fat people's diets were based on the climate effects of 'fatty foods like meat and dairy products', which they assumed fat people disproportionately ate. 'Greenhouse gas benefits of fighting obesity', p.300.
64. Edwards and Roberts, 'Population adiposity and climate change', p.2.
65. Ibid., p.3.
66. An unscientific but nevertheless interesting study of what different BMIs can actually look like is at http://kate harding.net/bmi-illustrated/
67. www.sciencenews.org/view/generic/id/6307/stepping_off_the_scale.
68. Raymond C Browning and Rodger Kram, 'Energetic Cost

and Preferred Speed of Walking in Obese vs. Normal Weight Women', *Obesity Research* 13, (2005) pp.891-99.

69. Edwards and Roberts, 'Population adiposity and climate change', p.4.
70. Michaelowa and Dransfeld, 'Greenhouse gas benefits of fighting obesity', p.306.
71. Ibid.
72. *Livestock's Long Shadow*, Livestock, Environment and Development Initiative (LEAD) and UN Food and Agriculture Organisation (FAO), (Rome 2006).
73. Ibid., pp.270-5
74. Ibid., p.17
75. Ibid., p.15
76. Ibid., p.283
77. Jessica Bellarby, Bente Foereid, Ashley Hastings and Pete Smith, *Cool Farming: Climate Impacts of Agriculture and Mitigation Potential*, (Greenpeace 2008), p.5.
78. *Food Matters*, p.8.
79. Garnett, *Cooking up a Storm*, p.3.
80. *Zero Carbon Britain 2030: A New Energy Strategy. The Second Report of the Zero Carbon Britain Project*, (Centre for Alternative Technology 2010), p.194.
81. *Livestock's Long Shadow*, p.xxi
82. *The Independent*, 10th December 2006.
83. www.alternet.org, 11th August 2008.
84. Monbiot, *Heat*, p.146. The FAO admitted in 2010 that it had underestimated the contribution of the transport system to climate change, and that taking road building, car manufacturing and so on into account, transport's share of global emissions was closer to 26%. http://www.theweek.co.uk/politics/climate-change/15721/meat-eaters-arent-causing-climate-change-un-admits
85. Garnett, *Cooking up a Storm*, pp.49-52.
86. Ibid., p.25.

87. Bellarby et al., *Cool Farming,* pp.23-34.
88. See for example Steven Shrubman, *Trade, Agriculture and Climate Change: How Agricultural Trade Policies Fuel Climate Change,* (Institute for Agriculture and Trade Policy, Minneapolis 2000), accessed at www.iatp.org; and Andy Jones, *Eating Oil. Food Supply in a Changing Climate,* (Sustain and Elm Farm Research Centre, 2001).
89. Shrubman, *Trade, Agriculture and Climate Change,* p.12.
90. Jones, *Eating Oil,* p.10.
91. Ibid., p.7.
92. Ibid, p.20.
93. Caroline Lucas, *Stopping the Great Food Swap. Relocalising Europe's Food Supply,* The Greens/European Free Alliance in the European Parliament, (2001), p.vi.
94. Jones, *Eating Oil,* p.17.
95. Felicity Lawrence, *Not on the Label. What really goes into the food on your plate,* (Penguin, London 2004), p.87.
96. Jones, *Eating Oil,* pp.76-8.
97. Eshel and Martin, 'Diet, Energy and Global Warming', p.15.
98. *The Independent,* 10th December 2006, http://www.independent.co.uk/environment/climate-change/cow-emissions-more-damaging-to-planet-than-co2-from-cars-427843.html
99. *The Independent,* 5th February 2012, http://www.independent.co.uk/environment/climate-change/meat-trade-emissions-equal-to-half-of-all-britains-cars-6423173.html. To be fair to *The Independent,* the headline-grabbing hook was in the abstract of the article, 'This is equivalent to a 50% reduction in current exhaust-pipe emissions from the entire UK passenger car fleet. Hence realistic choices about diet can make substantial differences to embodied GHG emissions', Berners-Lee et al., 'Greenhouse gas impacts of realistic dietary choices', abstract, p.184.
100. http://www.guardian.co.uk/media/mind-your-language/

2010/may/17/mind-your-language-david-marsh
101. Garnett, *Cooking up a Storm*, p.34.
102. Ibid., p.45.
103. Christopher L Weber and H Scott Matthews, 'Food Miles and the Relative Climate Impacts of Food Choices in the United States', *Environmental Science and Technology*, vol.42, no.10, (2008), pp.3508-13, p.3512.
104. Berners-Lee et al., 'Greenhouse gas impacts of realistic dietary choices', pp.184-90.
105. Shrubman, *Trade, Agriculture and Climate Change*, p.10.
106. Jones, *Eating Oil*, pp.76-8.
107. Ibid., pp.52-3.
108. Colin Hines, *Localization. A Global Manifesto*, (Earthscan, London and Sterling VA, 2000), p.141.
109. Naomi Klein, *No Logo 10*, http://www.naomiklein.org/articles/2009/11/revisiting-no-logo-ten-years-later, and see Chris Nineham, 'Anti-capitalism ten years after Seattle' for a useful commentary on the anti-capitalist movement, http://www.counterfire.org/index.php/theory/54-anti-capitalism/3550-anti-capitalism-ten-years-after-seattle.
110. Lawrence, *Not on the Label*, p.xiv. She makes a point of distancing herself from these arguments, commenting that globalisation could not be resisted 'any more than the weavers could resist the Industrial Revolution'.
111. Alexander Cockburn and Jeffrey St Clair, *Five Days that Shook the World. Seattle and Beyond*, (Verso, New York 2000), p.28.
112. Ibid., p.20.
113. Ray Kiely, *The Clash of Globalisations. Neo-Liberalism, the Third Way and Anti-Globalisation*, (Brill, Leiden-Boston 2005), p.222.
114. Jones, *Eating Oil*, p.68
115. Garnett, *Cooking up a Storm*, p.111.
116. Bellarby et al., *Cool Farming*, p.36.

117. Gold, *Global Benefits of Eating Less Meat*, p.64.
118. 'McCartney urges "meat-free" days to tackle climate change', *The Independent*, 15th June 2009.
119. Thomas, *Stuffed*, p.4.
120. Bob Holmes, 'What's the beef with meat?' *New Scientist*, 17th July 2010, pp.28-31, p.28.
121. Berners-Lee et al., 'Greenhouse gas impacts of realistic dietary choices', p.189.
122. Neale, *Stop Global Warming*, p.16
123. Bellarby et al., *Cool Farming*, p.25.
124. Eshel and Martin, 'Diet, Energy and Global Warming', p.12.
125. *Zero Carbon Britain*, pp.208-10.
126. *Meat Consumption. Trends and environmental implications*, Food Ethics Council, Report of the Business Forum Meeting, (20th November 2007), p.2, accessed at http://www.foodethicscouncil.org/files/business-forum201107.pdf
127. Dauvergne, *Shadows of Consumption*, p.139.
128. *Livestock's Long Shadow*, p.16.
129. Eschel and Martin, 'Diet, Energy and Global Warming', p.4.
130. http://www.1010uk.org/people#how_can_we, 1st September 2009.
131. Tim Jackson, *Motivating Sustainable Consumption. A review of evidence on consumer behaviour and behavioural change*. A report to the Sustainable Development Research Network, (Centre for Environmental Strategy, University of Surrey 2005), pp.106-9. On changing consumer behaviour, also see Inge Røpke, 'The dynamics of willingness to consume', *Ecological Economics*, 28 (1999), pp.399-420; Christer Sanne, 'Willing consumers or locked in? Policies for a sustainable consumption', *Ecological Economics* 42 (2002), pp.273-87; and *Weathercocks and Signposts* (2008).
132. Gold, *Global Benefits of Eating Less Meat*, pp.63-4.
133. *Zero Carbon Britain*, p.211.

134. See for example Kenneth F Kiple, *A Moveable Feast. Ten Millennia of Food Globalization,* (Cambridge UP, Cambridge 2007), pp.295-6.
135. John Beffes and Tasios Haniotis, *Placing the 2006/8 Commodity Price Boom into Perspective,* World Bank Policy Research Working Paper 5371, (July 2010), p.10.
136. Emiko Ohnuki-Tierney, 'McDonald's in Japan: Changing Manners and Etiquette', *Golden Arches East,* James L Watson (ed), (Stanford UP, Stanford 1997), pp.161-82, p.167.
137. Mille R Creighton, 'The Depato: Merchandising the West while selling Japaneseness', *Remade in Japan. Everyday Life and Consumer Taste in a Changing Society,* Joseph J Tobin (ed.), (Yale UP, New Haven & London 1992), pp.42-57, p.46.
138. See for example Reinhold Wagnleitner, *Cola-Colonization and the Cold War. The Cultural Mission of the United States in Austria after the Second World War,* trans. Diana M Wolf, (U of North Carolina P 1994).
139. Thomas L Friedman, *The Lexus and the Olive Tree,* (HarperCollins, New York 1999), p.195. The statement is no longer true: it was contradicted shortly after publication by Nato's bombing of Serbia, and subsequently by Israel's invasion of Lebanon in 2006 and the 2008 war between Russia and Georgia. Friedman argued in the 2000 edition of the book that the bombing of Belgrade actually proved his point, as he maintained that the Serbians' desire to return to the global order symbolised by McDonald's helped to bring the war to a conclusion, (New York 2000), pp.252-3.
140. Ibid (1999), p.309.
141. Eric B Ross, 'Patterns of Diet and Forces of Production. An Economic and Ecological History of the Ascendancy of Beef in the United States Diet', *Beyond the Myths of Culture. Essays on Cultural Materialism,* Eric B Ross (ed.), (Academic Press, New York 1980), pp.181-225.
142. *Food Matters,* p.3.

143. Ibid, p.5.
144. Ibid., p.38.
145. Ibid., p.15.
146. Ibid., p.36.
147. http://actonco2.direct.gov.uk/actonco2/home/campaigns/drive-5-miles-less-a-week.html, accessed 1st February 2010.
148. Lawrence, *Not on the Label*, p.78.
149. Jones, *Eating Oil*, pp.62-3.
150. Strategy Unit, *Food Matters*, p.18.
151. Ibid., p.37.
152. *Zero Carbon Britain*, p.219.
153. *Global Benefits of Eating Less Meat*, p.10.
154. Pimental et al., 'Reducing Energy Inputs in the US Food System', p.468.
155. *Global Benefits of Eating Less Meat*, p.63
156. Ibid., p.20.
157. See chapter 1, p.50.
158. www.meatlessmonday.com/about, accessed 25th June 2009.
159. www.meatlessmonday.com, accessed 25th June 2009 and 19th March 2012. The site layout had changed in the intervening period, but the focus on obesity, as opposed to just not eating meat, had stayed the same.
160. Esme Choonara and Sadie Robinson, *Hunger in a World of Plenty: What's Behind the Global Food Crisis?*, (Socialist Workers Party 2008), p.1.
161. Donald Mitchell, *A Note on Rising Food Prices*, World Bank Policy Research Working Paper 4682, (July 2008), p.2.
162. Tim Jones, *The Great Hunger Lottery. How banking speculation causes food crises*, (World Development Movement, July 2010), p.5.
163. The FAO estimated 923 million people in total were malnourished in 2007, including the 75 million increase in that year. See *The State of Food Insecurity in the World 2008. High food prices and food security – threats and opportunities*,

(FAO 2008), p.2.

164. For a partial list of food price protests, with links to fuller reports, see Loren Peabody, 'Rising Food Prices, Rising Rood Protests', www.foodfirst.org/en/node/2086, 11th April 2008.
165. 'Cry out in anger at Egypt's show trials', *Socialist Worker*, 12th July 2008.
166. Ian Black, 'Struggling country where bread means life', *The Guardian*, 12th April 2008.
167. http://www.bbc.co.uk/news/business/market_data/commodities/158426/twelve_month.stm.
168. http://www.guardian.co.uk/commentisfree/2010/aug/13/food-prices-past-crises
169. Alex Preston, 'Lessons from the wheat crisis', *New Statesman*, 13th August 2010.
170. Peabody, 'Rising Food Prices', www.foodfirst.org/en/node/2086.
171. *Food Matters*, p.27.
172. http://www.euractiv.com/en/cap/energy-prices-speculation-blame-recent-food-price-hike-says-world-bank-news-497158.
173. Mitchell, *Note on Rising Food Prices*, pp.16-17. See also Brian Tokar, 'Biofuels and the Global Food Crisis', Fred Magdoff and Brian Tokar (eds.), *Agriculture and Food in Crisis. Conflict, Resistance and Renewal*, (Monthly Review Press, New York 2010), pp.121-38, p.121.
174. Baffes and Haniotis, *2006-2008 Commodity Price Boom*.
175. Jones, *The Great Hunger Lottery*, pp.8-9.
176. Choonara and Robinson, *Hunger in a World of Plenty*, p.14.
177. Jones, *The Great Hunger Lottery*, p.21.
178. Ibid., p.9.
179. *Food Matters*, p.27.
180. Jones, *The Great Hunger Lottery*, p.10.
181. Joachim von Braun, *The World Food Situation. New Driving Forces and Required Actions*, (International Food Policy

Research Institute, Washington DC, December 2007), p.5.

182. Andrew Boswell, '2008: The year to stop agrofuels', http://www.biofuelwatch.org.uk/greenworld08.php.

183. Dr Rachel Smolker, Brian Tokar, Anne Petermann and Eva Hernandez, *The Real Cost of Agrofuels: Food, Forest and the Climate*, (Global Forest Coalition and Global Justice Ecology Project, 2007), p.4.

184. See for example *Agrofuels: Towards a Reality Check in Nine Key Areas*, (Biofuelwatch et al., June 2007), pp.13-16.

185. Ben Block, 'UK Biofuels sources are largely unknown', *WorldWatch Institute*, 15th August 2008, http://www.world-watch.org/node/5861.

186. *Agrofuels: Reality Check*, pp.21-3.

187. *Zero Carbon Britain*, p.193.

188. Ibid., p.220.

189. *Livestock's Long Shadow*, p.283

190. Garnett, *Cooking up a Storm*, p.148.

191. *Thinking about the Future of Food. The Chatham House Food Supply Scenarios*, (Chatham House Food Supply Project, May 2008), quotation at p.4.

192. Blake Alcott, 'The Sufficiency Strategy: Would Rich World Frugality Lower Environmental Impact?', *Ecological Economics* 64 (2008), pp.770-86, p.782.

193. Tim Jackson, *Prosperity without Growth? The Transition to a Sustainable Economy*, (Sustainable Development Commission 2009), p.5.

194. *Livestock's Long Shadow*, p.283.

195. Dauvergne, *Shadows of Consumption*, p.4. For further discussion of population, see chapter 4.

196. *New York Times*, 7th April 2008.

197. *The Guardian*, 30th May 2008, http://www.guardian.co.uk /environment/2008/may/30/food.china1

198. Jonathan A Foley et al., 'Solutions for a Cultivated Planet', *Nature* 478, 20 October 2011, pp.337-42, p.337.

199. *Livestock's Long Shadow*, pp.125-76.
200. Gold, *Global Benefits of Eating Less Meat*, Jonathan Porritt, *Forward*, p.5.
201. *Livestock's Long Shadow*, p.133-35. The calculation of the water used to produce beef is therefore essentially the total rainfall on the soil on which the animal feed was grown, so exaggerating the water required for beef rather than plant food, Simon Fairlie, *Meat. A Benign Extravagance*, (Permanent Publications, Vermont 2010), pp.63-68.
202. Gold, *Global Benefits of Eating Less Meat*, Jonathan Porritt, *Forward*, p.5.
203. Gold, *Global Benefits of Eating Less Meat*, p.30.
204. Fairlie, *Meat*, particularly pp.157-87.
205. Gold, *Global Benefits of Eating Less Meat*, p.30. See for example Herman E Daly, *Beyond Growth. The Economics of Sustainable Development*, (Beacon Press, Boston 1996), p.5 for an early use of the concept: 'Add to that [the effects of 1.2 billion Chinese attaining Western standards of living] the ecological consequences from agriculture when the Chinese begin to eat higher on the food chain'.
206. http://www.guardian.co.uk/environment/2007/jan/27/usnews.frontpagenews
207. See for example Brett Clark and Richard York, 'Rifts and Shifts. Getting to the Root of Environmental Crises', *Monthly Review* 60, no.6, (2008), pp.13-24; Victor Wallis, 'Capitalist and Socialist Responses to the Ecological Crisis', ibid., pp.25-40; and Jackson, *Prosperity without Growth?*, pp.48-57.
208. *Zero Carbon Britain*, p.219.
209. Gold, *Global Benefits of Eating Less Meat*, p.34.
210. Princen, 'Consumption and Environment', p.348.
211. John Kenneth Galbraith, *The Affluent Society*, (1958), 3rd ed., (Houghton Mifflin Company, Boston 1976).
212. Ibid., p.xxiii. These areas of Appalachia were the focus of Lyndon Johnson's War on Poverty in 1964, and it was

suggested that poverty here was seen by the government as more serious than elsewhere in the US not because this was the poorest area, but because it was largely white. Whether or not this also underlay Galbraith's concern is difficult to say. See Bill Bryson, *The Lost Continent. Travels in Small Town America*, (Abacus, London 1990), pp.102-3.

213. Galbraith, *Affluent Society*, p.224.
214. For a discussion of J S Mill's theory and its use in modern steady state thinking, see the Center for the Advancement of the Steady State Economy, www.steadystate.org, accessed 11th May 2009.
215. For a relatively concise explanation of the theory, see Brian Czech and Herman E Daly, 'In My Opinion: The Steady State Economy – What It Is, What It Entails and Connotes', *Wildlife Society Bulletin* 32 (2) (2004), pp.598-605.
216. Herman E Daly, 'From Empty World Economics to Full World Economics: Recognising a Historical Turning Point in Economic Development', *Population, Technology and Lifestyle* (1992), pp.23-37, p.25.
217. Andrew Simms, *Ecological Debt. The Health of the Planet and the Wealth of Nations*, (Pluto Press, London and Ann Arbor, Michigan, 2005).
218. Czech and Daly, 'In My Opinion', p.602. Czech also states elsewhere that the proposition of development without growth is not socialist. Brian Czech, 'The Foundation of a New Conservation Movement. Professional Society Positions on Economic Growth, *Bioscience* 57 (1) (2007), p.6, accessed at www.biosciencemag.org.
219. Jackson, *Prosperity without Growth?*, p.46.
220. See for example Dauvergne, *Shadows of Consumption*, (2008) or Jackson, *Prosperity without Growth?*, (2009).
221. See for example Marx, *Grundrisse*, pp.89-96.
222. Galbraith, *Affluent Society*, p.62.
223. Ibid., p.127.

224. Ibid., pp.122-4.
225. Interview in *The Catholic Herald*, 22nd December 1978, accessed at http://www.margaretthatcher.org/document/103793 and quoted in Owen Jones, *Chavs. The Demonization of the Working Class*, (Verso, London 2011), p.64.
226. Galbraith, *Affluent Society*, p.2.
227. Karl Marx, *Capital. A Critique of Political Economy*, (1893), 3 vols., (Foreign Languages Publishing House, Moscow 1957), vol.2, p.410.
228. Rosa Luxemburg, *Einfuhrung in die Nationalokononien*, (Berlin 1925), p.275, cited in Ernest Mandel, *Late Capitalism*, trans. Joris De Bres, (Verso, London and New York 1978), p.150.
229. Marx, *Grundrisse*. pp.419-20.
230. Marx, *Capital*, vol.2, p.403.
231. Ibid.
232. See chapter 1, pp.26.
233. Marx, *Grundrisse*, p.285.
234. Ibid., p.286.
235. Ibid., p.287.
236. Ibid., pp.122-7.
237. *Weathercocks and Signposts*, pp.14-23.
238. Paul Stern, 'Toward a working definition of consumption for environmental research and policy', Paul C Stern et al. (eds.), *Environmentally Significant Consumption: Research Directions*, (National Academy Press, Washington DC 1997), pp.12-25, p.16.
239. Ibid., p.20.
240. Ibid., p.22.
241. Princen, 'Consumption and environment', p.350.
242. Ibid., p.357.
243. Ibid., p.358.
244. Ibid., p.359.
245. Ibid.

246. Alcott, 'The sufficiency strategy', p.771.
247. Ibid., p.780.
248. Røpke, 'The dynamics of willingness to consume', pp.403-5.
249. Dauvergne, *Shadows of Consumption*, p.14.
250. Simms, *Ecological Debt*, p.157.
251. See for example Selina Todd, *The People. The Rise and Fall of the Working Class 1910-2010*, (John Murray, London 2014), pp.131-132, who points out that rationing meant that many working-class families could afford to consume more meat and dairy products in 1943 than they had in the 1930s.
252. Jackson, *Prosperity without Growth?*, p.7.
253. Clive Hamilton, *Growth Fetish*, (Pluto, London 2004), pp.92-5.
254. Garnett, *Cooking Up a Storm*, pp.111-112.
255. Ibid., p.54.
256. See for example H L Bedes, 'The Historical Context of the Essay on Population', *Introduction to Malthus*, D V Glass (ed.), (Frank Cass, London 1953), pp.3-24, p.22.
257. There were seven editions of the *Principle of Population* in total, six during Malthus' lifetime and a seventh shortly after his death. The revisions in the 3rd – 7th editions largely amend the presentation rather than the content of the argument, although the 3rd edition also removed some controversial passages from the earlier editions (see below).
258. T R Malthus, O*n the Principle of Population*, 7th ed., (1834), (Everyman), 2 vols., vol.1, pp.5-11.
259. Andrew R B Ferguson, 'Malthus over a 270 Year Perspective', *Optimum Population Trust*, vol.8, no.1, (2008), pp.20-23, p.20.
260. Bedes, 'Historical Context of the Essay on Population', p.22.
261. Yves Charbit, *Economic, Social and Demographic Thought in the XIXth Century: The Population Debate from Malthus to Marx*, (Springer, New York 2009), p.5. On the neo-Malthusians, see in particular Allan Chase, *The Legacy of*

Malthus. The Social Costs of the New Scientific Racism, (U of Illinois P, Chicago 1975) and Fred Pearce, *Peoplequake. Mass Migration, Ageing Nations and the Coming Population Crash*, (Eden Project, London 2010), pp.45-106.

262. John F Rohe, *A Bicentennial Malthusian Essay. Conservation, Population and the Indifference to Limits*, (Rhodes & Easton, Michigan 1997), p.9.
263. See chapter 2, pp.65.
264. David Pimental, 'Ecological Systems, Natural Resources and Food Supplies', *Food, Energy and Society*, David and Marcia Pimental (eds.), 2nd ed., (University of Colorado 1996), pp.23-40, p.40.
265. Robert Goodland, 'The Case that the World has Reached its Limits', *Population, Technology and Lifestyle. The Transition to Sustainability*, eds. Robert Goodland, Herman E Daly and Salah El Serafy, (Island Press, Washington DC and Coveto, California, 1992), pp.3-22, p.7, following the original calculations by PM Vitousek, Paul Ehrlich, AN Ehrlich and PA Mason, *Bioscience* 36, 368, (1986).
266. www.npg.org/pospapers/nogrowth.html
267. www.steadystate.org/CASSEPositiononEG.html
268. http://steadystate.org/discover/organizations-that-support-steady-state-principles
269. Charbit, *Population Debate*, pp.39-41. What Malthus was imagining was the hardest-working workers clawing their way up the ladder, not an increase in wages for all; he thought that would be counterproductive, as money in workers' pockets would act as a disincentive to hard work.
270. Pimental, 'Ecological Systems', p.40.
271. See chapter 2, p.65.
272. Malthus, *Principle of Population*, p.6.
273. Ibid., p.10.
274. Ibid., p.11.
275. Daly, *Beyond Growth*, p.14.

276. Malthus, *Principle of Population*, p.3.
277. Ibid.
278. William Cobbett, *Rural Rides*, (1830), ed. George Woodcock, (Penguin, London 1967), p.317.
279. As made for example by Arthur Jensen, 'How much can we boost IQ and scholastic achievement?' *Harvard Educational Review* 33, (1969), pp.1-33 and Richard J Herrnstein and Charles Murray, *The Bell Curve: The reshaping of American life by difference in intelligence*, (Simon & Schuster, New York 1994). For an effective history and an elegant refutation of this racist rubbish, see Stephen Jay Gould, *The Mismeasure of Man*, 2nd ed, (W. W. Norton, London and New York 1996).
280. As advanced for example by Richard Dawkins, *The Selfish Gene*, (Oxford UP, Oxford 1976) and many subsequent works. Fruitful sources of counter-arguments include Steven Rose, Leon J Kamin and R C Lewontin, *Not In Our Genes. Biology, Ideology and Human Nature*, (Pantheon, London and New York 1984); Richard Lewontin, *It Ain't Necessarily So. The Dream of the Human Genome and Other Illusions*, (Granta, London 2000); and Steven Rose, *The 21st Century Brain. Explaining, Mending and Manipulating the Mind*, 2nd ed., (Vintage, London 2006).
281. For recent discussions of arguments for the innateness of gender differences, see Deborah Cameron, *The Myth of Mars and Venus: Do men and women really speak different languages?*, (Oxford UP, Oxford 2007) and Cordelia Fine, *Delusions of Gender. The Real Science Behind Sex Differences*, (Icon Books, London 2010).
282. E O Wilson, *Sociobiology: The New Synthesis*, (Harvard UP, Cambridge, Mass. 1975), p.575.
283. Malthus, *Principle of Population*, p.6.
284. Malthus, *Principle of Population*, 2nd ed., (1803), p.531.
285. Frederick Engels, *The Condition of the Working Class in England* (1892), (Penguin, London 1969), p.308.

286. Gould, *Mismeasure of Man*, p.29.
287. Howard Zinn, *A People's History of the United States, 1492-present*, 3rd ed., (Routledge, London 2003), pp.377-87.
288. Gould, *Mismeasure of Man*, pp.31-2. It's important not to get carried away with identifying every work of right-wing determinism with periods of reaction, but it is worth noting that the publication of E O Wilson's landmark *Sociobiology* in 1975 coincided with what Zinn identifies as a systematic reaction against the protest movements of the late 1960s and early 1970s. See Zinn, *People's History of the United States*, pp.539-62.
289. Chase, *Legacy of Malthus*, p.72.
290. E.J. Hobsbawm and George Rudé, *Captain Swing*, (Lawrence & Wishart, London 1969), p.6.
291. Ibid.
292. Flora Thompson, *Lark Rise to Candleford*, (1945), (Penguin, London 2008), pp.79-80.
293. E.P.Thompson, *The Making of the English Working Class*, (Gollancz, New York 1963), p.221.
294. G.D.H.Cole and Raymond Postgate, *The Common People 1746-1946*, (Methuen, London 1961), p.276.
295. Ronald L Meek (ed.), *Marx and Engels on Malthus. Selections from the writings of Marx and Engels dealing with the theories of Thomas Robert Malthus*, (People's Pub. House, London 1953), introduction, p.17.
296. T R Malthus, *Principles of Political Economy considered with a view to their practical application, 2nd edition with considerable editions from the author's own manuscript and an original memoir*, (Basil Blackwell, Oxford 1951), pp.xv-xvi.
297. Marx, *Capital*, vol.1, pp.628-40; Frederick Engels, 'The Myth of Overpopulation', *Outlines of a Critique of Political Economy* (1844), reprinted in Meek, *Marx and Engels on Malthus*, pp.57-60; Engels, *Condition of the Working Class*, pp.112-13.
298. Cobbett, *Rural Rides*, p.317.

299. Paul Ehrlich, *The Population Bomb*, (Sierra Club, New York 1969), p.1. For a useful discussion of Ehrlich and his ilk, see Pearce, *Peoplequake*, pp.66-8.
300. Lindsey Grant, *Juggernaut. Growth on a Finite Planet*, (Seven Locks Press, Santa Ana, California 1996), p.2.
301. See for example Rohe, *Bicentennial Malthusian Essay*, p.49; Ronald D Lee, 'The Second Tragedy of the Commons', Kingsley Davis and Mikhail S Bernstein (eds.), *Resources, Environment and Population: Present Knowledge, Future Growth*, Population and Development Review vol.16, (New York and Oxford 1991), pp.315-22, p.317.
302. Virginia Abernethy, 'Introduction', in Ester Boserup, *The Conditions of Agricultural Growth. The Economics of Agrarian Change under Population Pressure*, (1965), 2nd ed., (Transaction Publishers, New Brunswick and London 2006), pp.vii-xiii, p.ix.
303. http://www.theguardian.com/uk-news/2013/jul/16/rhetoric-reality-immigration?INTCMP=SRCH&guni=Article:in%20body%20link
304. http://www.theguardian.com/politics/2013/jul/25/greens-worried-high-immigration
305. Sandy Irvine, interview, 29th May 2014.
306. He was not alone in this; the photograph has been called 'the most influential environmental photograph ever taken', see http://www.abc.net.au/science/moon/earthrise.htm
307. Susan George, *How the Other Half Dies. The Real Reasons for World Hunger*, 2nd ed., (Penguin, London 1986), p.67.
308. Engels, *Condition of the Working Class in England*, p.309.
309. Engels, 'Myth of Overpopulation', p.58.
310. Ester Boserup, *The Conditions of Agricultural Growth. The Economics of Agrarian Change under Population Pressure*, (1965).
311. Nathan Key Fitz, 'Toward a Theory of Population-Development Interaction', Davis and Bernstein (eds.),

Resources, Environment and Population, pp.295-314, p.299.
312. Boserup, *Conditions of Agricultural Growth*, pp.119-21.
313. Emmanuel Le Roy Ladurie, *The Peasants of Languedoc*, (1966), trans. John Day, (U of Illinois P, Chicago/London 1974), p.311.
314. V.I.Lenin, 'The Working Class and Neo Malthusianism', *Pravda* 137, 16th June 1913, in *Lenin Collected Works*, (Progress Publishers, Moscow 1977), vol.19, pp.235-7, available from www.marxists.org.
315. Martin Empson, *Marxism and Ecology. Capitalism, Socialism and Future of the Planet*, (Socialist Worker 2009), p.19.
316. Goodland, 'The Case that the World has Reached its Limits', p.7.
317. *Thinking About the Future of Food*, Chatham House, p.5.
318. Pearce, *Peoplequake*, pp.84-90.
319. Eric B Ross, *The Malthus Factor. Population, Poverty and Politics in Capitalist Development*, (Zed Books, London 1998), p.199.
320. Pearce, *Peoplequake*, p.90.
321. Helena Norberg-Hodge, Todd Mannfield and Steven Gorelick, *Bringing the Food Economy Home. Local Alternatives to Global Agribusiness*, (Zed Books, London 2002), p.53.
322. Jason W Moore, 'Ecological Crises and the Agrarian Question in the World-Historical Perspective', *Monthly Review* 60, no.6, (November 2008), pp.54-63.
323. Engels, letter to Lange, 29th March 1865, Meek, *Marx and Engels on Malthus*, p.82.
324. Fairlie, *Meat. A Benign Extravagance*, pp.109-113.
325. Cited in David Arnold, *Famine. Social Crisis and Historical Change*, (Wiley-Blackwell, Oxford 1988), p.25. This view of the Black Death as a crisis caused by overpopulation has not gone away, as it is repeated in recent writing such as Jason W Moore, 'The Crisis of Feudalism, An Environmental History', *Organization and Environment* 15, 3(2002), pp.301-

22, p.306.

326. Cormac Ó Gráda, *Famine. A Short History,* (Princeton UP, Princeton and Oxford 2009), p.37.
327. Arnold, *Famine,* pp.54-5.
328. Amartya Sen, *Poverty and Famines. An Essay on Entitlement and Deprivation,* (Oxford UP, Oxford 1981), p.43. The Dutch famine of 1944 is another example of a famine which did affect most people, rather than just the poorest, although this of course cannot be regarded as a famine caused by Malthusian overpopulation. Even Malthus would have found it difficult to argue that the Second World War was an expression of Nature's need to reduce the population.
329. Ibid., pp.52-85
330. Ibid., p.166.
331. Ó Gráda, *Famine,* pp.160-90
332. John Newsinger, *The Blood Never Dried. A People's History of the British Empire,* (Bookmarks 2006), pp.34-8.
333. Brian Fagan, *The Little Ice Age. How Climate Made History 1300-1850,* (Basic Books, New York 2000), pp.181-94.
334. Newsinger, *The Blood Never Dried,* p.38.
335. Carlo Morelli, 'Behind the World Food Crisis', *International Socialism* 119 (2008), pp.37-49.
336. See for example Marge Piercy, *Body of Glass,* (Penguin, London-New York 1991), in which a lawless slum called The Glop covers most of the US from coast to coast.
337. J. Rockström et al., 'Planetary Boundaries: Exploring the Safe Operating Space for Humanity', *Ecology and Society* 14 (2), (2009): p.32.
338. Gerald H Haug et al., 'Climate and the Collapse of Maya Civilization', *Science* 299, (14 March 2003), pp.1731-5, p.1733. See also Jared Diamond, *Collapse, How Societies Choose to Fail or Survive,* (Penguin, London 2005), pp.157-77; and R B Gill, *The Great Maya Droughts: Water, Life and Death,* (U of New Mexico P, Albuquerque 2000).

339. Charles C Mann, *1491. New Revelations of the Americas before Columbus*, (Vintage, New York 2006), pp.308-12.
340. Foley et al., 'Solutions for a Cultivated Planet', p.337.
341. See for example *Livestock's Long Shadow*.
342. David Collett, 'Pastoralists and Wildlife: Image and Reality in Kenya Maasailand', David Anderson and Richard Grover (eds.), *Conservation in Africa. Peoples, Policies and Practice*, (Cambridge UP, Cambridge 1987), pp.129-48, p.138.
343. Katherine Harrowood and W A Rodgers, 'Pastoralism, Conservation and the Overgrazing Controversy', *Conservation in Africa*, pp.111-28, p.115.
344. Tristram Stuart, *Waste: Uncovering the Global Food Scandal*, (Penguin, London 2009), p.xvi.
345. Tom Quested and Hannah Johnson, *Household Food and Drink Waste in the UK*, (WRAP 2009), pp.25-7; Ashok Chapagain and Keith James, *The water and carbon footprint of household food and drink waste in the UK*, (WRAP and WWF, March 2011), p.4.
346. *Achieving Food Security*, Commission on Food Security and Climate Change, p.3.
347. Quested and Johnson, *Household Food and Drink*, p.6.
348. Stuart, *Waste*, p.xix.
349. Andrew Simms and Caroline Lucas MP, *The New Home Front. Showing Leadership: How we can learn from Britain's war time past in an age of dangerous climate change and energy insecurity*, (Green Party, 2011).
350. Lawrence, *Not on the Label*, p.78.
351. Stuart, *Waste*, pp.48-9.
352. *Private Eye*, 'Ad Nauseam', 24th January–6th February 2014, p.13.
353. http://www.dailymail.co.uk/news/article-1357741/in-court-charged-theft-finding-woman-took-food-tesco-bin.html
354. Jeff Shantz, *One Person's Garbage... Another Person's Treasure: Dumpster Driving, Freeganism and Anarchy*, http://verb.lib.

lehigh.edu/index.php/verb/article/viewFile/19/19. For more resources on freeganism, see http://freegan.info.

355. Personal interview, 21st October 2013.
356. See http://www.prsc.org.uk/
357. For details of the campaign, see http://notesco.wordpress.com/
358. http://www.thewi.org.uk/standard.aspx?id=10926
359. Stuart, *Waste*, p.10.
360. http://www.guardian.co.uk/environment/2008/jul/07/food.waste1; and http://www.telegraph.co.uk/news/politics/labour/2259954/Gordon-Brown-Stop-wasting-food.html
361. http://www.sustainweb.org/news.php?id=221; and http://www.telegraph.co.uk/news/uknews/2264012/Food-waste-Consumers-tell-Gordon-Brown-to-bogof-if-he-wants-to-ban-buy-one-get-one-free.html
362. Stuart, *Waste*, p.78.
363. http://notbuyinganything.blogspot.com/2011/03/reducing-food-waste.html.
364. As for example in the UK government's 2007 *Waste Strategy for England*, http://www.tsoshop.co.uk/bookstore.asp?Action=Book&ProductId=9780215543233
365. http://www.eatwashington.com/article/chew_on_this_us_food_waste_boosting_global_warming.
366. Roberts and Edwards, *Energy Glut*, p.136.
367. Stuart, *Waste*, p.70.
368. Roberts and Edwards, *Energy Glut*, pp.134-5.
369. Julie Hill, *The Secret Life of Stuff. A Manual for a New Material World*, (Vintage, London 2011), p.213.
370. Professor Tim Lang, quoted in Stuart, *Waste*, p.70.
371. Ibid., p.121.
372. Hill, *Secret Life of Stuff*, pp.276-84.
373. John Bellamy Foster, Brett Clark and Richard York, *The Ecological Rift. Capitalism's War on the Earth*, (Monthly Review Press, New York 2010), p.132.

374. D J Peterson, *Troubled Lands. The Legacy of Soviet Environmental Destruction*, (Westview Press, Boulder 1993), p.10.
375. Bellamy Foster et al., *Ecological Rift*, pp.133-5.
376. See Tony Cliff, *State Capitalism in Russia*, (Pluto, London 1988) for discussion of the proper identification of the Soviet system.
377. John Bellamy Foster et al., *Ecological Rift*, p.204.
378. As suggested in the context of reducing climate change from industry by the authors of *The Hartwell Paper. A new direction for climate policy after the crash of 2009*, (Institute for Science, Innovation and Society, University of Oxford and London School of Economics 2010).
379. John Bellamy Foster et al., *Ecological Rift*, p.37.
380. Carolyn Wyman, *Better Than Homemade. Amazing foods that changed the way we eat*, (Quirk Books, Philadelphia 2004), p.9.
381. Laura Shapiro, *Something from the Oven. Reinventing dinner in 1950s America*, (Penguin, New York 2004), pp.43-51.
382. Betty Friedan, *The Feminine Mystique*, (1963), (Penguin, London 1982), pp.182-8.
383. Quoted in Shapiro, *Something from the Oven*, p.64.
384. See chapter 1, pp.XX.
385. John Bellamy Foster et al., *Ecological Rift*, pp.209-10.
386. Richard Lewontin and Richard Lewins, *Biology under the Influence*, (Monthly Review Press, New York 2007), pp.321-8.
387. Peter Rosset, 'Fixing our Global Food System: Food Sovereignty and Redistributive Land Reform', Magdoff and Tokar (eds.), *Agriculture and Food in Crisis* (Monthly Review Press, New York 2010), pp.189-205, p.200.
388. Ibid., p.201.
389. http://stevesurbangarden.blogspot.co.uk/, accessed 12th May 2014.
390. http://windsorcsa.blogspot.co.uk/, accessed 12th May 2014.
391. http://windsorcsa.blogspot.co.uk/2013/08/new-article-on-

ourwindsorca.html, accessed 12th May 2014.

392. See Lewontin and Lewins, *Biology under the Influence*, pp.343-64.
393. Miguel A. Altieri et al., 'The greening of the "barrios": Urban agriculture for food security in Cuba', *Agriculture and Human Values*, 16 (1999), pp.131-2.
394. Miguel A. Altieri and Fernando R. Funes-Monzote, 'The Paradox of Cuban Agriculture', *Monthly Review* (January 2012), http://monthlyreview.org/2012/01/01/the-paradox-of-cuban-agriculture
395. Dennis T Avery, 'Cubans starve on a diet of lies', 2nd April 2009, http://www.cgfi.org/2009/04/cubans-starve-on-diet-of-lies-by-dennis-t-avery/
396. Fernando Funes, Miguel Altieri and Peter Rosset, 'The Avery Diet: The Hudson Institute's misinformation campaign against Cuban agriculture', May 2009, http://globalalternatives.org/files/AveryCubaDiet.pdf.
397. See for example http://www.theguardian.com/global-development/poverty-matters/2013/sep/26/venezuela-food-shortages-rich-country-cia; and BBC 'Venezuela Out of Stock', *Crossing Continents*, 5th September 2013.
398. Lee Brown, personal communication, 5th October 2013.
399. www.facebook.com/seanhawkey65/media_set?set=a.10153314691855118.1073741828.639095117&type=1, accessed 15th May 2014.
400. http://www.theguardian.com/world/2014/apr/01/venezuela-food-shortage-id-cards
401. http://www.theguardian.com/global-development/poverty-matters/2013/sep/26/venezuela-food-shortages-rich-country-cia
402. Jules Pretty, 'Can Ecological Agriculture Feed Nine Billion People?', Magdoff and Tokar (eds.), *Agriculture and Food in Crisis*, pp.283-98, p.296.
403. For an excellent discussion of the proletarianisation of the

farmer, see Lewontin and Lewins, *Biology under the Influence*, pp.329-41.

404. On the importance of totality, see Georg Lukács, *History and Class Consciousness. Studies in Marxist Dialectics*, (1922), trans. Rodney Livingstone, (The Merlin Press, London 1974), pp.1-26.

405. http://seedstock.com/2013/03/07/oregon-urban-farm-collective-brings-neighbors-together-transforexchanges-food-for-hours-inspire-other-urban-communities/, accessed 12th May 2014.

406. A version of part of this section appears as Elaine Graham-Leigh, 'The green movement in Britain', Matthias Dietz and Heiko Garrelts, *Routledge Handbook of the Climate Change Movement*, (Routledge, London and New York 2014), pp.107-16.

407. http://www.guardian.co.uk/environment/2013/feb/28/public-concern-environment?CMP=twt_gu

408. This tendency is not restricted to the UK: the snowstorm which hit the eastern US in October 2011 was similarly treated by climate change sceptics as evidence that global warming theories were flawed. http://www.bishop-hill.net/blog/2011/10/31/snow-in-new-england.html. For research on the connection between particular weather events and concern about climate change, see Simon D Donner and Jeremy McDaniels, 'The influence of natural temperature fluctuations on opinions about climate change in the US since 1990', *Climatic Change*, 5th February 2013, http://link.springer.com/article/10.1007/s10584-012-0690-3

409. Ted Nordhaus and Michael Shellenberger, 'The Death of Environmentalism. Global Warming Politics in a Post-Environmental World', (2004), www.breakthrough.org; and *Break Through. Why We Can't Leave Saving The Planet To Environmentalists*, (2007), 2nd ed., (Mariner Books, New York 2009).

410. Mark Lynas, *The God Species. How the Planet Can Survive the Age of Humans*, (Fourth Estate, London 2011), p.10.
411. See the range of articles on nuclear power reposted on www.monbiot.com.
412. George Monbiot and Chris Goodall, 'The moral case for nuclear power', http://www.monbiot.com/2011/08/08/the-moral-case-for-nuclear-power/.
413. Ibid.
414. Lynas, *The God Species*, p.12.
415. Nordhaus and Shellenberger, *Break Through*, p.271.
416. Ibid., p.viii.
417. Lynas, *The God Species*, p.66.
418. Anthony Giddens, *The Politics of Climate Change*, (Polity Press, Cambridge 2009).
419. http://eprints.lse.ac.uk/27939/
420. George Monbiot, 'Nuked by friend and foe', http://www.monbiot.com/2009/02/20/nuked-by-friend-and-foe
421. See for example George Monbiot, ''The lost world', http://www.monbiot.com/2011/05/02/the-lost-world
422. From a list of 271 local wind farm campaigns on http://www.countryguardian.net/Campaign%20Windfarm%20Action%20Groups.htm, accessed 14th November 2011.
423. My thoughts on the wind-up of Climate Camp owe much to Sophie Lewis' excellent *The rise and fall (and rise) of the Camp for Climate Action UK: notes on interventions in challenging carbon democracy*, unpublished ms, (Oxford UP, Oxford, 2011). However, my conclusions and any misapprehensions are my own. Also interesting on this point are comments made by some of the interviewees in the 2011 film *Just Do It* on their thoughts on the climate change movement post Copenhagen: http://justdoitfilm.com/.
424. William Morris, *News from Nowhere, or An Epoch of Rest, being some chapters from a utopian romance*, (1890), ed. David Leopold, (Oxford UP, Oxford 2003), p.13.

425. Ibid., p.87.
426. Ibid.
427. Edward Bellamy, *Looking Backwards from 2000 to 1887*, (1888), (Project Gutenberg 2008), chapter 13, http://www.gutenberg.org/files/624/624-h/624-h.htm#chap13
428. Ibid., chapter 14.
429. Ursula Le Guin, *The Dispossessed*, (1974), (Gollancz, London 1999), p.207.
430. Ibid., p.260.
431. Ursula K Le Guin, *Always Coming Home*, (1985), (Gollancz, London 1988), p.438.
432. Ibid., p.443.

Bibliography

Achieving Food Security in the face of Climate Change, Commission on Food Security and Climate Change (2011)

Agrofuels: Towards a Reality Check in Nine Key Areas, (Biofuelwatch et al., June 2007)

Alcott, Blake, 'The Sufficiency Strategy: Would Rich World Frugality Lower Environmental Impact?', *Ecological Economics* 64 (2008), pp.770-86

Altieri, Miguel A, and Fernando R. Funes-Monzote, 'The Paradox of Cuban Agriculture', *Monthly Review* (January 2012), http://monthlyreview.org/2012/01/01/the-paradox-of-cuban-agriculture

Altieri, Miguel A, et al., 'The greening of the "barrios": Urban agriculture for food security in Cuba', *Agriculture and Human Values*, 16 (1999), pp.131-2

Arnold, David, *Famine. Social Crisis and Historical Change*, (Wiley-Blackwell, Oxford 1988)

Bedes, H L,'The Historical Context of the Essay on Population', *Introduction to Malthus*, D V Glass (ed.), (Frank Cass, London 1953), pp.3-24

Beffes, John and Tasios Haniotis, *Placing the 2006/8 Commodity Price Boom into Perspective*, World Bank Policy Research Working Paper 5371, (July 2010)

Bellamy Foster, John, Brett Clark and Richard York, *The Ecological Rift. Capitalism's War on the Earth*, (Monthly Review Press, New York 2010)

Bellamy, Edward, *Looking Backwards from 2000 to 1887*, (1888), (Project Gutenberg 2008)

Bellarby, Jessica, Bente Foereid, Ashley Hastings and Pete Smith, *Cool Farming: Climate Impacts of Agriculture and Mitigation Potential*, (Greenpeace 2008)

Berners-Lee, M, C. Hoolohan, H. Cammack and C.N.Hewitt,

'The relative greenhouse gas impacts of realistic dietary choices', *Energy Policy* 43, (April 2012), pp.184-90

Boserup, Ester, *The Conditions of Agricultural Growth. The Economics of Agrarian Change under Population Pressure,* (1965), 2nd ed., (Transaction Publishers, New Brunswick and London 2006)

Bouchard, C and S N Blair, 'Introductory comments for the consensus on physical activity and obesity', *Medicine and Science in Sports and Exercise* 31, 11, section 4, pp.98-501

Browning, Raymond C and Rodger Kram, 'Energetic Cost and Preferred Speed of Walking in Obese vs. Normal Weight Women', *Obesity Research* 13, (2005) pp.891-9

Bryson, Bill, *The Lost Continent. Travels in Small Town America,* (Abacus, London 1990)

Cameron, Deborah, *The Myth of Mars and Venus: Do men and women really speak different languages?,* (Oxford UP, Oxford 2007)

Campos, Paul, *The Obesity Myth. Why America's Obsession with Weight is Hazardous to your Health,* (Gotham Books, New York 2004)

Chapagain, Ashok, and Keith James, *The water and carbon footprint of household food and drink waste in the UK,* (WRAP and WWF, March 2011)

Charbit, Yves, *Economic, Social and Demographic Thought in the XIXth Century: The Population Debate from Malthus to Marx,* (Springer, New York 2009)

Chase, Allan, *The Legacy of Malthus. The Social Costs of the New Scientific Racism,* (U of Illinois P, Chicago 1975)

Choonara, Esme and Sadie Robinson, *Hunger in a World of Plenty: What's Behind the Global Food Crisis?,* (Socialist Workers Party 2008)

Clark, Brett and Richard York, 'Rifts and Shifts. Getting to the Root of Environmental Crises', *Monthly Review* 60, no.6, (2008), pp.13-24

Cliff, Tony, *State Capitalism in Russia,* (Pluto, London 1988)

Cobbett, William, *Rural Rides,* (1830), ed. George Woodcock, (Penguin, London 1967)

Cockburn, Alexander and Jeffrey St Clair, *Five Days that Shook the World. Seattle and Beyond,* (Verso, New York 2000)

Cole, G D H, and Raymond Postgate, *The Common People 1746-1946,* (Methuen, London 1961)

Collett, David, 'Pastoralists and Wildlife: Image and Reality in Kenya Maasailand', David Anderson and Richard Grover (eds.), *Conservation in Africa. Peoples, Policies and Practice,* (Cambridge UP, Cambridge 1987), pp.129-48

Creighton, Mille R, 'The Depato: Merchandising the West while selling Japaneseness', *Remade in Japan. Everyday Life and Consumer Taste in a Changing Society,* Joseph J Tobin (ed.), (Yale UP, New Haven & London 1992), pp.42-57

Czech, Brian and Herman E Daly, 'In My Opinion: The Steady State Economy – What It Is, What It Entails and Connotes', *Wildlife Society Bulletin* 32 (2) (2004)

Czech, Brian, 'The Foundation of a New Conservation Movement. Professional Society Positions on Economic Growth', *Bioscience* 57 (1) (2007), p.6, accessed at www.biosciencemag.org

Daly, Herman E, 'From Empty World Economics to Full World Economics: Recognising a Historical Turning Point in Economic Development', *Population, Technology and Lifestyle* (1992), pp.23-37

Daly, Herman E, *Beyond Growth. The Economics of Sustainable Development,* (Beacon Press, Boston 1996)

Dauvergne, Peter, *The Shadows of Consumption. Consequences for the Global Environment,* (MIT, Cambridge, Mass. and London 2008)

Davies, Nick, *Flat Earth News. An Award-winning Reporter Exposes Falsehood, Distortion and Propaganda in the Global Media,* (Chatto & Windus, London 2008)

Dawkins, Richard, *The Selfish Gene*, (Oxford UP, Oxford 1976)

Diamond, Jared, *Collapse, How Societies Choose to Fail or Survive*, (Penguin, London 2005)

Edwards, Phil and Ian Roberts, 'Population Adiposity and Climate Change', *International Journal of Epidemiology*, (2009), pp.1-4.

—, 'Transport policy is food policy', *The Lancet*, vol 371, no.9639, (17th May 2008)

Ehrlich, Paul, *The Population Bomb*, (Sierra Club, New York 1969)

Empson, Martin, *Marxism and Ecology. Capitalism, Socialism and Future of the Planet*, (Socialist Worker 2009)

Engels, Frederick, 'The Myth of Overpopulation', *Outlines of a Critique of Political Economy* (1844), reprinted in Ronald L Meek (ed.), *Marx and Engels on Malthus. Selections from the writings of Marx and Engels dealing with the theories of Thomas Robert Malthus*, (People's Pub. House, London 1953)

—, *The Condition of the Working Class in England* (1892), (Penguin, London 1969)

Eshel, Gideon and Pamela Martin, 'Diet, Energy and Global Warming', *Earth Interactions* 10, (March 2006), pp.1-17

Fagan, Brian, *The Little Ice Age. How Climate Made History 1300-1850*, (Basic Books, New York 2000)

Fairlie, Simon, *Meat. A Benign Extravagance*, (Permanent Publications, Vermont 2010)

Ferguson, Andrew R B, 'Malthus over a 270 Year Perspective', *Optimum Population Trust*, vol.8, no.1, (2008), pp.20-23

Fine, Cordelia, *Delusions of Gender. The Real Science Behind Sex Differences*, (Icon Books, London 2010)

Foley, Jonathan A, et al., 'Solutions for a Cultivated Planet', *Nature* 478, 20 October 2011, pp.337-42

Food Matters. Towards and Strategy for the 21st Century, The Strategy Unit, (2008)

Foresight - Tackling Obesities: Future Choice - Project Report, Government Office for Science (2007), available at www.foresight.gov.uk

Friedan, Betty, *The Feminine Mystique*, (1963), (Penguin, London 1982)

Friedman, Thomas L, *The Lexus and the Olive Tree*, (HarperCollins, New York 1999)

Galbraith, John Kenneth, *The Affluent Society*, (1958), 3rd ed., (Houghton Mifflin Company, Boston 1976)

Gard, Michael and Jan Wright, *The Obesity Epidemic. Science, Morality and Ideology*, (Routledge, London and New York 2005)

Garnett, Tara, *Cooking up a Storm. Food, Greenhouse Gas Emissions and our Changing Climate*, (Food Climate Research Network, Centre for Environmental Strategy, University of Surrey, 2008)

George, Susan, *How the Other Half Dies. The Real Reasons for World Hunger*, 2nd ed., (Penguin, London 1986)

Giddens, Anthony, *The Politics of Climate Change*, (Polity Press, Cambridge 2009)

Gill, R B, *The Great Maya Droughts: Water, Life and Death*, (U of New Mexico P, Albuquerque 2000).

Gold, Mark, *The Global Benefits of Eating Less Meat*, (Compassion in World Farming Trust 2004)

Goodall, Chris, *How to Live a Low-Carbon Life. The Individual's Guide to Stopping Climate Change*, (Earthscan, London-Sterling VA 2007)

Goodland, Robert, 'The Case that the World has Reached its Limits', *Population, Technology and Lifestyle. The Transition to Sustainability*, eds. Robert Goodland, Herman E Daly and Salah El Serafy, (Island Press, Washington DC and Coveto, California, 1992), pp.3-22

Gould, Stephen J, *The Mismeasure of Man*, 2nd ed, (W. W. Norton, London and New York 1996).

Graham-Leigh, Elaine, 'The green movement in Britain', Matthias Dietz and Heiko Garrelts, *Routledge Handbook of the*

Climate Change Movement, (Routledge, London and New York 2014), pp.107-16

Grant, Lindsey, *Juggernaut. Growth on a Finite Planet*, (Seven Locks Press, Santa Ana, California 1996)

Hamilton, Clive, *Growth Fetish*, (Pluto, London 2004)

Haroon Akram-Lodhi, A, *Hungry for Change. Farmers, Food Justice and the Agrarian Question*, (Fernwood Publishing, Halifax and Winnipeg 2013)

Harrowood, Katherine, and W A Rodgers, 'Pastoralism, Conservation and the Overgrazing Controversy', *Conservation in Africa*, pp.111-28

Haug, Gerald H, et al., 'Climate and the Collapse of Maya Civilization', *Science* 299, (14 March 2003), pp.1731-5

Healthy Weight, Healthy Lives. A Cross-Government Strategy for England, (Cross-Government Obesity Unit, January 2008).

Herrnstein, Richard J and Charles Murray, *The Bell Curve: The reshaping of American life by difference in intelligence*, (Simon & Schuster, New York 1994)

Hill, Julie, *The Secret Life of Stuff. A Manual for a New Material World*, (Vintage, London 2011)

Hines, Colin, *Localization. A Global Manifesto*, (Earthscan, London and Sterling VA, 2000)

Hobsbawm, E J, and George Rudé, *Captain Swing*, (Lawrence & Wishart, London 1969)

Jackson, Tim, *Motivating Sustainable Consumption. A review of evidence on consumer behaviour and behavioural change*. A report to the Sustainable Development Research Network, (Centre for Environmental Strategy, University of Surrey 2005)

—, *Prosperity without Growth? The Transition to a Sustainable Economy*, (Sustainable Development Commission 2009)

Jensen, Arthur, 'How much can we boost IQ and scholastic achievement?' *Harvard Educational Review* 33, (1969), pp.1-33

Jones, Andy, *Eating Oil. Food Supply in a Changing Climate*, (Sustain and Elm Farm Research Centre, 2001).

Jones, Owen, *Chavs. The Demonization of the Working Class,* (Verso, London 2011)

Jones, Tim, *The Great Hunger Lottery. How banking speculation causes food crises,* (World Development Movement, July 2010)

Key Fitz, Nathan, 'Toward a Theory of Population-Development Interaction', Kingsley Davis and Mikhail S Bernstein (eds.), *Resources, Environment and Population: Present Knowledge, Future Growth,* Population and Development Review vol.16, (New York and Oxford 1991), pp.295-314

Kiely, Ray, *The Clash of Globalisations. Neo-Liberalism, the Third Way and Anti-Globalisation,* (Brill, Leiden-Boston 2005)

Kiple, Kenneth F, *A Moveable Feast. Ten Millennia of Food Globalization,* (Cambridge UP, Cambridge 2007)

Lawrence, Felicity, *Not on the Label. What really goes into the food on your plate,* (Penguin, London 2004)

Le Guin, Ursula K, *Always Coming Home,* (1985), (Gollancz, London 1988)

—, *The Dispossessed,* (1974), (Gollancz, London 1999)

Le Roy Ladurie, Emmanuel, *The Peasants of Languedoc,* (1966), trans. John Day, (U of Illinois P, Chicago/London 1974)

Lee, Ronald D, 'The Second Tragedy of the Commons', Kingsley Davis and Mikhail S Bernstein (eds.), *Resources, Environment and Population: Present Knowledge, Future Growth,* Population and Development Review vol.16, (New York and Oxford 1991), pp.315-22

Lenin, V I, 'The Working Class and Neo Malthusianism', *Pravda* 137, 16th June 1913, in *Lenin Collected Works,* (Moscow 1977), vol.19, pp.235-7, available from www.marxists.org

Lewis, Sophie, *The rise and fall (and rise) of the Camp for Climate Action UK : notes on interventions in challenging carbon democracy,* unpublished ms, (Oxford UP, Oxford 2011)

Lewontin, Richard, and Richard Lewins, *Biology under the Influence,* (Monthly Review Press, New York 2007)

Lewontin, Richard, *It Ain't Necessarily So. The Dream of the Human*

Genome and Other Illusions, (Granta, London 2000)

Livestock's Long Shadow, Livestock, Environment and Development Initiative (LEAD) and UN Food and Agriculture Organisation (FAO), (Rome 2006)

Lucas, Caroline, *Stopping the Great Food Swap. Relocalising Europe's Food Supply,* The Greens/European Free Alliance in the European Parliament, (2001)

Lukács, Georg, *History and Class Consciousness. Studies in Marxist Dialectics,* (1922), trans. Rodney Livingstone, (The Merlin Press, London 1974)

Lynas, Mark, *The God Species. How the Planet Can Survive the Age of Humans,* (Fourth Estate, London 2011)

Malthus, T R, *On the Principle of Population,* 7th ed., (1834), (Everyman), 2 vols.

—, *Principles of Political Economy considered with a view to their practical application, 2nd edition with considerable editions from the author's own manuscript and an original memoir,* (Basil Blackwell, Oxford 1951)

Mandel, Ernest, *Late Capitalism,* trans. Joris De Bres, (Verso, London and New York 1978)

Mann, Charles C, *1491. New Revelations of the Americas before Columbus,* (Vintage, New York 2006)

Marx, Karl, *Capital. A Critique of Political Economy,* (1893), 3 vols., (Foreign Languages Publishing House, Moscow 1957)

—, *Grundrisse. Foundations of the Critique of Political Economy (Rough Draft),* (1939) trans. Martin Nicolaus, (Penguin/New Left Review, London 1973)

Meat Consumption. Trends and environmental implications, Food Ethics Council, Report of the Business Forum Meeting, (20th November 2007) accessed at http://www.foodethics council.org/files/businessforum2011 07.pdf

Meek, Ronald L, (ed.), *Marx and Engels on Malthus. Selections from the writings of Marx and Engels dealing with the theories of Thomas Robert Malthus,* (People's Pub. House, London 1953)

Michaelowa, Axel, and Björn Dransfield, 'Greenhouse gas benefits of fighting obesity', *Ecological Economics* 66, (2008)

Mitchell, Donald, *A Note on Rising Food Prices*, World Bank Policy Research Working Paper 4682, (July 2008)

Monbiot, George, *Heat. How to Stop the Planet Burning*, (Allen Lane, London 2006).

Moore, Jason W, 'Ecological Crises and the Agrarian Question in the World-Historical Perspective', *Monthly Review* 60, no.6, (November 2008)

—, 'The Crisis of Feudalism, An Environmental History', *Organization and Environment* 15, 3(2002), pp.301-22

Morelli, Carlo, 'Behind the World Food Crisis', *International Socialism* 119 (2008)

Morris, William, *News from Nowhere, or An Epoch of Rest, being some chapters from a utopian romance*, (1890), ed. David Leopold, (Oxford UP, Oxford 2003)

Neale, Jonathan, *Stop Global Warming. Change the World*, (Bookmarks Publications, London 2008).

Newsinger, John, *The Blood Never Dried. A People's History of the British Empire*, (Bookmarks 2006)

Norberg-Hodge, Helena, Todd Mannfield and Steven Gorelick, *Bringing the Food Economy Home. Local Alternatives to Global Agribusiness*, (Zed Books, London 2002)

Nordhaus, Ted, and Michael Shellenberger, 'The Death of Environmentalism. Global Warming Politics in a Post-Environmental World', (2004), www.breakthrough.org

—, *Break Through. Why We Can't Leave Saving The Planet To Environmentalists*, (2007), 2nd ed., (Mariner Books, New York 2009)

Ó Gráda, Cormac, *Famine. A Short History*, (Princeton UP, Princeton and Oxford 2009)

Ohnuki-Tierney, Emiko, 'McDonald's in Japan: Changing Manners and Etiquette', *Golden Arches East*, James L Watson (ed), (Stanford UP, Stanford 1997), pp.161-82

Patel, Raj, *Stuffed and Starved. Markets, Power and the Hidden Battle for the World Food System*, (Portobello Books, London 2007)

Pearce, Fred, *Peoplequake. Mass Migration, Ageing Nations and the Coming Population Crash*, (Eden Project, London 2010)

Peterson, D J, *Troubled Lands. The Legacy of Soviet Environmental Destruction*, (Westview Press, Boulder 1993)

Piercy, Marge, *Body of Glass*, (Penguin, London-New York 1991)

Pimental, D, 'Reducing Energy Inputs in the US Food System', *Human Ecology* 36, no.4 (August 2008), pp.459-71

—, 'Ecological Systems, Natural Resources and Food Supplies', *Food, Energy and Society*, David and Marcia Pimental (eds.), 2nd ed., (U of Colorado P, Colorado 1996), pp.23-40

Pretty, Jules, 'Can Ecological Agriculture Feed Nine Billion People?', Fred Magdoff and Brian Tokar (eds.), *Agriculture and Food in Crisis. Conflict, Resistance and Renewal*, (Monthly Review Press, New York 2010), pp.283-98

Princen, Thomas, 'Consumption and Environment: Some Conceptual Issues', *Ecological Economics* 31 (1999)

Quested, Tom, and Hannah Johnson, *Household Food and Drink Waste in the UK*, (WRAP 2009)

Roberts, Ian with Phil Edwards, *The Energy Glut. Climate Change and the Politics of Fatness*, (Zed Books, London and New York 2010)

Rockström, J, et al., 'Planetary Boundaries: Exploring the Safe Operating Space for Humanity', *Ecology and Society* 14 (2), (2009): p.32.

Rohe, John F, *A Bicentennial Malthusian Essay. Conservation, Population and the Indifference to Limits*, (Rhodes & Easton, Michigan 1997)

Røpke, Inge, 'The dynamics of willingness to consume', *Ecological Economics*, 28 (1999), pp.399-420

Rose, Steven, Leon J Kamin and R C Lewontin, *Not In Our Genes. Biology, Ideology and Human Nature*, (Pantheon, London and New York 1984)

Rose, Steven, *The 21st Century Brain. Explaining, Mending and Manipulating the Mind*, 2nd ed., (Vintage, London 2006).

Ross, Eric B, 'Patterns of Diet and Forces of Production. An Economic and Ecological History of the Ascendancy of Beef in the United States Diet', *Beyond the Myths of Culture. Essays on Cultural Materialism*, Eric B Ross (ed.), (Academic Press, New York 1980), pp.181-225.

Ross, Eric B, *The Malthus Factor. Population, Poverty and Politics in Capitalist Development*, (Zed Books, London 1998)

Rosset, Peter, 'Fixing our Global Food System: Food Sovereignty and Redistributive Land Reform', Fred Magdoff and Brian Tokar (eds.), *Agriculture and Food in Crisis. Conflict, Resistance and Renewal*, (Monthly Review Press, New York 2010), pp.189-205

Sanne, Christer, 'Willing consumers or locked in? Policies for a sustainable consumption', *Ecological Economics* 42 (2002), pp.273-87

Sen, Amartya, *Poverty and Famines. An Essay on Entitlement and Deprivation*, (Oxford UP, Oxford 1981

Shapiro, Laura, *Something from the Oven. Reinventing dinner in 1950s America*, (Penguin, New York 2004)

Shrubman, Steve, *Trade, Agriculture and Climate Change: How Agricultural Trade Policies Fuel Climate Change*, (Institute for Agriculture and Trade Policy, Minneapolis 2000), accessed at www.iatp.org

Simms, Andrew, and Caroline Lucas MP, *The New Home Front. Showing Leadership: How we can learn from Britain's war time past in an age of dangerous climate change and energy insecurity*, (Green Party, 2011)

Simms, Andrew, *Ecological Debt. The Health of the Planet and the Wealth of Nations*, (Pluto Press, London and Ann Arbor, Michigan, 2005).

Smolker, Dr Rachel, Brian Tokar, Anne Petermann and Eva Hernandez, *The Real Cost of Agrofuels: Food, Forest and the*

Climate, (Global Forest Coalition and Global Justice Ecology Project, 2007)

Stern, Paul, 'Toward a working definition of consumption for environmental research and policy', Paul C Stern, Thomas Dietz, Vernon W Rutton, Robert H Socolow and James L Sweeney (eds.), *Environmentally Significant Consumption: Research Directions*, (National Academy Press, Washington DC 1997), pp.12-25

Stuart, Tristram, *Waste: Uncovering the Global Food Scandal*, (Penguin, London 2009)

The Hartwell Paper. A new direction for climate policy after the crash of 2009, (Institute for Science, Innovation and Society, University of Oxford and London School of Economics 2010)

The State of Food Insecurity in the World 2008. High food prices and food security – threats and opportunities, (FAO 2008)

Thinking about the Future of Food. The Chatham House Food Supply Scenarios, (Chatham House Food Supply Project, May 2008)

Thomas, Pat, *Stuffed. Positive Action to Prevent a Global Food Crisis*, (Soil Association 2010)

Thompson, E P, *The Making of the English Working Class*, (Gollancz, New York 1963)

Thompson, Flora, *Lark Rise to Candleford*, (1945), (Penguin, London 2008)

Todd, Selina, *The People. The Rise and Fall of the Working Class 1910-2010*, (John Murray, London 2014)

Tokar, Brian, 'Biofuels and the Global Food Crisis', Fred Magdoff and Brian Tokar (eds.), *Agriculture and Food in Crisis. Conflict, Resistance and Renewal*, (Monthly Review Press, New York 2010), pp.121-38

Trevino, Roberto P, et al., 'Diabetes risk, low fitness and energy insufficiency levels among children from poor families', *Journal of the American Dietetic Association*, 108 no.11, (November 2008), pp.1846-53

Tudge, Colin, *Feeding People is Easy*, (Pari, Italy 2007)

Vitousek, P M, Paul Ehrlich, AN Ehrlich and PA Mason, *Bioscience* 36, 368, (1986)

von Braun, Joachim, *The World Food Situation. New Driving Forces and Required Actions,* (International Food Policy Research Institute, Washington DC, December 2007)

Wagnleitner, Reinhold, *Cola-Colonization and the Cold War. The Cultural Mission of the United States in Austria after the Second World War,* trans. Diana M Wolf, (U of North Carolina P 1994)

Wallis, Victor, 'Capitalist and Socialist Responses to the Ecological Crisis', *Monthly Review* 60, no.6, (2008), pp.25-40

Weathercocks and Signposts. The environmental movement at a cross-roads, (World Wildlife Fund, April 2008).

Weber, Christopher L, and H Scott Matthews, 'Food Miles and the Relative Climate Impacts of Food Choices in the United States', *Environmental Science and Technology,* vol.42, no.10, (2008), pp.3508-13

Weighty Matters. The London Findings of the National Child Measurement Programme 2006-2008, London Health Observatory, (May 2009)

White, Rachel, 'Undesirable Consequences? Resignifying Discursive Constructions of Fatness in the Obesity "Epidemic"', Corinna Tomrley and Ann Kaloski-Naylor (eds.), *Fat Studies in the UK,* (Raw Nerve Books, York 2009), pp.69-81.

Wilson, E O, *Sociobiology: The New Synthesis,* (Harvard UP, Cambridge, Mass. 1975)

Wyman, Carolyn, *Better Than Homemade. Amazing foods that changed the way we eat,* (Quirk Books, Philadelphia 2004)

Zero Carbon Britain 2030: A New Energy Strategy. The Second Report of the Zero Carbon Britain Project, (Centre for Alternative Technology 2010)

Zinn, Howard, *A People's History of the United States, 1492-present,* 3rd ed., (Routledge, London 2003) zero books

Contemporary culture has eliminated both the concept of the public and the figure of the intellectual. Former public spaces – both physical and cultural – are now either derelict or colonized by advertising. A cretinous anti-intellectualism presides, cheerled by expensively educated hacks in the pay of multinational corporations who reassure their bored readers that there is no need to rouse themselves from their interpassive stupor. The informal censorship internalized and propagated by the cultural workers of late capitalism generates a banal conformity that the propaganda chiefs of Stalinism could only ever have dreamt of imposing. Zer0 Books knows that another kind of discourse – intellectual without being academic, popular without being populist – is not only possible: it is already flourishing, in the regions beyond the striplit malls of so-called mass media and the neurotically bureaucratic halls of the academy. Zer0 is committed to the idea of publishing as a making public of the intellectual. It is convinced that in the unthinking, blandly consensual culture in which we live, critical and engaged theoretical reflection is more important than ever before.

Weird Realism Lovecraft and Philosophy
Graham Harman
As Hölderlin was to Martin Heidegger and Mallarmé to Jacques Derrida, so is H.P. Lovecraft to the Speculative Realist philosophers.
Paperback: September 28, 2012 978-1-78099-252-5 $24.95 £14.99.
eBook: September 28, 2012 978-1-78099-907-4 $9.99 £6.99.

Sweetening the Pill or How We Got Hooked on Hormonal Birth Control
Holly Grigg-Spall
Is it really true? Has contraception liberated or oppressed women?
Paperback: September 27, 2013 978-1-78099-607-3 $22.95 £12.99.
eBook: September 27, 2013 978-1-78099-608-0 $9.99 £6.99.

Why Are We The Good Guys? Reclaiming Your Mind From The Delusions Of Propaganda
David Cromwell
A provocative challenge to the standard ideology that Western power is a benevolent force in the world.
Paperback: September 28, 2012 978-1-78099-365-2 $26.95 £15.99.
eBook: September 28, 2012 978-1-78099-366-9 $9.99 £6.99.

The Truth about Art Reclaiming quality
Patrick Doorly
The book traces the multiple meanings of art to their various sources, and equips the reader to choose between them.
Paperback: August 30, 2013 978-1-78099-841-1 $32.95 £19.99.

Bells and Whistles More Speculative Realism
Graham Harman
In this diverse collection of sixteen essays, lectures, and interviews Graham Harman lucidly explains the principles of Speculative Realism, including his own object-oriented philosophy.

Paperback: November 29, 2013 978-1-78279-038-9 $26.95 £15.99.
eBook: November 29, 2013 978-1-78279-037-2 $9.99 £6.99.

Towards Speculative Realism: Essays and Lectures Essays and Lectures
Graham Harman
These writings chart Harman's rise from Chicago sportswriter to co founder of one of Europe's most promising philosophical movements: Speculative Realism.
Paperback: November 26, 2010 978-1-84694-394-2 $16.95 £9.99.
eBook: January 1, 1970 978-1-84694-603-5 $9.99 £6.99.

Meat Market Female flesh under capitalism
Laurie Penny
A feminist dissection of women's bodies as the fleshy fulcrum of capitalist cannibalism, whereby women are both consumers and consumed.
Paperback: April 29, 2011 978-1-84694-521-2 $12.95 £6.99.
eBook: May 21, 2012 978-1-84694-782-7 $9.99 £6.99.

Translating Anarchy The Anarchism of Occupy Wall Street
Mark Bray
An insider's account of the anarchists who ignited Occupy Wall Street.
Paperback: September 27, 2013 978-1-78279-126-3 $26.95 £15.99.
eBook: September 27, 2013 978-1-78279-125-6 $6.99 £4.99.

One Dimensional Woman
Nina Power
Exposes the dark heart of contemporary cultural life by examining pornography, consumer capitalism and the ideology of women's work.
Paperback: November 27, 2009 978-1-84694-241-9 $14.95 £7.99.
eBook: July 1, 2012 978-1-78099-737-7 $9.99 £6.99.

Dead Man Working

Carl Cederstrom, Peter Fleming

An analysis of the dead man working and the way in which capital is now colonizing life itself.

Paperback: May 25, 2012 978-1-78099-156-6 $14.95 £9.99.

eBook: June 27, 2012 978-1-78099-157-3 $9.99 £6.99.

Unpatriotic History of the Second World War

James Heartfield

The Second World War was not the Good War of legend. James Heartfield explains that both Allies and Axis powers fought for the same goals - territory, markets and natural resources.

Paperback: September 28, 2012 978-1-78099-378-2 $42.95 £23.99.

eBook: September 28, 2012 978-1-78099-379-9 $9.99 £6.99.

Find more titles at www.zero-books.net